BRITISH CHANTERELLES AND TOOTH-FUNGI

An account of the British Cantharelloid and stipitate Hydnoid Fungi

D. N. PEGLER

P. J. ROBERTS

B. M. SPOONER

Royal Botanic Gardens, Kew

First published 1997

Cover Design by Jeff Eden for Media Resources, Royal Botanic Gardens, Kew

ISBN 1 900347 15 6

Printed in the European Union by Continental Printing, Belgium

CONTENTS

INTRODUCTION

The Chanterelles are amongst the most popular of all the wild, edible mushrooms. They are often the first to be sought after by the amateur collector, and are generally very conspicuous, with striking colours, often bright orange or yellow, and easily spotted on the woodland floor. The Tooth Fungi are less common, but when they do occur they can be locally abundant, and some are also striking to look at, having either bright violaceous pigments, or a striking black and white appearance.

When Carlton Rea published his account of the *British Basidiomycetae* (1922), the Chanterelles, or *Cantharellaceae*, were included as the only family of the *Cantharellineae* within the order *Agaricales*, along with typical mushrooms and toadstools. They were separated from the *Agaricineae* by virtue of the hymenium 'spread over the surface of narrow, obtuse gills or folds, or quite smooth'. All of the Spine or Tooth-Fungi with stipes were gathered together within the order *Aphyllophorales* because the hymenium was 'exposed from the first, amphigenous or unilateral, and spread over teeth'. Most were included in the *Hydnaceae* and were almost entirely confined to the genus *Hydnum* L.: Fr., which included thirty-three British taxa. Such a broad arrangement, often referred to as the Friesian classification after the Swedish mycologist, Elias Fries, was based upon the gross morphological appearance of the fruitbodies and survived for more than a century.

A more 'natural' system of classification, which linked related species and separated unrelated species, was not universally accepted until the publication of two major monographs, the first on the '*Cantharelloid Fungi*' (1966) by E.J.H. Corner, and the other, '*Die terrestrischen Stachelpilze Europas*' (1975) by R.A. Maas Geesteranus. Both authors concentrated their attention on a hyphal analysis of the fruitbodies, defining specialized hyphal types arranged in recognizable hyphal systems. Their results were combined with analyses of the chemical properties of the fruitbodies and especially the micro-chemistry of the basidiospores.

Today, the *Cantharellaceae* are placed in an order, the *Cantharellales*, which is far removed from the *Agaricales*, and also includes a newly defined *Hydnaceae*, restricted to *Hydnum repandum* L.: Fr. and its allies. The remaining stipitate, hydnoid fungi are included in a number of genera, such as *Auriscalpium* Gray, *Bankera* Coker & Beers ex Pouzar, *Hericium* Pers., *Hydnellum* P. Karst, *Mucronella* Fr., *Phellodon* P. Karst. and *Sarcodon* P. Karst., which may be found in two, unrelated orders, the *Hericiales* and the *Thelephorales*.

All of the *Cantharellales* are edible and some of the *Hericiales*, particularly the Monkey Head (*Hericium erinaceum* (Bull.) Pers.), are highly prized and sometimes commercially cultivated, both for their esculent and medicinal properties. The Scaly Tooth (*Sarcodon imbricatus* (L.: Fr.) P. Karst.) is sometimes eaten, but the remaining Tooth-Fungi tend to be too tough or bitter.

In this book, the British species are brought together in a modern, revised account for the first time. Each species is provided with a full description of both the macroscopical and microscopical characters, a full synonymy, and details on habitat, distribution and frequency. Each species is illustrated in colour, together with line drawings of the micro-characters. The account of the *Cantharellales* has been provided by D.N. Pegler, the *Hericiales* and *Bankeraceae* by P.J. Roberts, and the hydnoid *Thelephoraceae* by B.M. Spooner. The definitions used for the higher taxa and in the keys are based upon the British species, and groups from elsewhere in the World may require wider definitions.

Acknowledgements

The authors are grateful to the following for kindly providing photographs for use or consideration: A.M. Ainsworth, A.W. Brand, G. Dickson, E.E. Emmett, S. Evans, N.W. Legon, P. Marstad, S. Ryman, Å. Strid.

WHAT ARE CHANTERELLES AND TOOTH-FUNGI ?

The Chanterelles and Tooth-Fungi are a small group of mostly mushroom-like, fleshy fungi which have often been grouped together but, in reality, represent a number of genera dispersed through several separate families and three orders. For the most part, their structure is simple, showing minimal development of cystidial elements in the hymenium, an uncomplicated hyphal construction and pileal surface (pileipellis), and gymnocarpic development, lacking any protective veil layers in the developmental stages. They are regarded by many as 'primitive Higher Fungi' and, in some cases, as relicts, suggesting ancestral forms for many of the basidiomycete families and orders.

In addition to the radially lamellate hymenophore, the mushrooms, or agarics (*Agaricales*), are separated from the rest of the homobasidiomycetes by virtue of their hyphal construction. Their hyphal system is monomitic and the generative-hyphae tend to swell or inflate, thereby producing the well-known, speedy 'mushroom growth' resulting from the rapid uptake of water. The hyphae of most aphyllophoroid basidiomycetes do not inflate, and growth is much slower. Inflated hyphae may be found, however, in all the Chanterelles (*Cantharellaceae* and *Craterellaceae*), the Wood Urchins (*Hydnaceae*), the Spine Fungi (*Hericiaceae*), the Spiny Cap-Fungi (*Sarcodon*) and the Drab and Violet Tooth-Fungi (*Bankera*). These fungi have an essentially soft-fleshy texture and are relatively short-lived. The fruitbodies of the remaining Tooth-Fungi, namely the Corky Spine Fungi (*Hydnellum*) and the Scented Spine Fungi (*Phellodon*), are formed of generative-hyphae which do not inflate, growth is slower and they can often survive for several months. Maas Geesteranus (1962) considered in detail the hyphal construction found in hydnoid fungi.

The hymenium of the Chanterelles is an unusual and a major characteristic. It thickens continually so that new basidia are continually added and grow over and beyond the old basidia. There is no clearly defined subhymenial layer producing a palisade of basidia, as is the case in the *Agaricales* and all the Tooth-Fungi. This results in the hymenophore, which is originally smooth, bulging outwards in such a way as to become slightly veined, wrinkled or radially ridged. The ridges run from the stipe to the pileal margin. Corner (1966: 5) defines such development as the 'folds are lines of excessive intercalary growth of the initially smooth hymenophore'. The ridges are obtuse and their edges are always fertile, i.e. they bear basidia from the beginning. Although in a number of Chanterelles these ridges can resemble the lamellae found in the fruitbodies of agarics, the resemblance is superficial. In agarics, the hymenophore is produced by radial plates of sterile tissue growing downwards from the underside of the pileus, to form the hymenophoral, or hymenium-bearing, trama. A subhymenial layer is produced over the surface of the tramal plates, which gives rise to a simple palisade of basidia. The basidia are formed simultaneously. This is the normal hymenial type in most homobasidiomycetes, and is described as a 'eu-hymenium', unlike the 'thickening hymenium' of the Chanterelles. In the Chanterelles there are no vertically orientated tramal plates, but the smooth hymenophore wrinkles and thereby pulls down the hyphae from the underside of the pileus. A section through a hymenophoral ridge therefore does not reveal a vertical arrangement of hyphal growth as in an agaric lamella, but hyphae which are loosely and irregularly arranged. Similarly, the subhymenial layer is indistinct and loosely differentiated.

The second major characteristic of the Chanterelles is the nature of the basidia. The basidium is very elongated, at times exceeding 100 µm, and often narrow and strongly tapering at the base. This is related to the thickening hymenium, through which the basidia need to push in order to grow beyond the upper, older basidia. In addition, the long basidium is cytologically characterized by the arrangement of the nuclear division at the time of meiosis. The meiotic

spindle is formed parallel to the longitudinal axis of the basidium. This is known as a stichobasidium, in contrast to the chiastobasidium of most agaricoid fungi, which produce a transverse meiotic spindle.

The fruitbody, or basidioma, of Chanterelles may be one of four types:

(i) ***Eucantharellus*-type.** The fruitbody is agaricoid and thick-fleshy, with a convex pileus which becomes flattened or depressed on expansion, a hymenophore which is 'falsely-lamellate', and a distinct, usually solid stipe e.g. *Cantharellus cibarius* Fr. (Fig. 1A)

(ii) ***Leptocantharellus*, plicati-type**. The pileus and stipe are well developed, but the stipe is hollow and the pileus frequently becomes perforated at the centre to produce an open cavity and the fruitbody is finally funnel-shaped. The hymenophore forms well developed 'false-lamellae' e.g. *Cantharellus tubiformis* Bull.: Fr. (Fig. 1B)

(iii) ***Leptocantharellus*, phlebini-type**. Similar to the plicato-type but the hymenophore is scarcely wrinkled remaining smooth or forming vein-like ridges e.g. *Cantharellus aurora* (Batsch) Kuyp. (Fig. 1C)

(iv) ***Craterellus*-type**. The stipe is only slightly developed or possibly not at all, but the pileus forms a deep funnel so that the lower surface appears on the outside of the funnel, and the upper pileal surface lines the funnel e.g. *Craterellus cornucopioides* (L.: Fr.) Pers. (Fig. 1D)

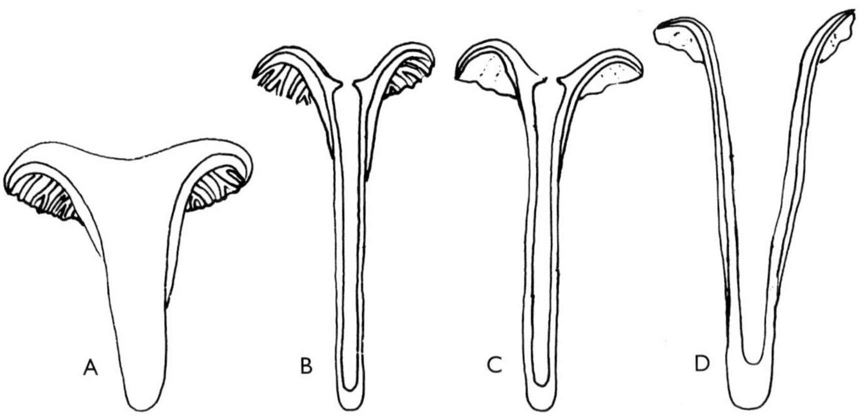

Fig. 1. Fruitbody types of Chanterelles. **A,** *Eucantharellus*-type; **B,** *Leptocantharellus*, plicati-type; **C,** *Leptocantharellus*, phlebini- type; **D,** *Craterellus*-type.

The Tooth-Fungi were originally considered as a single group of related species, based upon the hymenophore construction. The hymenophore is essentially composed of 'teeth', which are positively geotropic, that is to say they grow vertically downwards pointing towards the ground. The 'teeth', however, show a wide degree of variation. The Prickle Fungus (*Mucronella calva* (Alb. & Schwein.) Fr.) is composed of numerous, small, isolated spines, about 0.5–5 mm long, which are directly attached to the woody substratum. In the other genera, however, there is a fruitbody, although this can be much reduced in some of the species in the Spine Fungi. In the Monkey Head (*Hericium erinaceum* (Bull.) Pers.), the fruitbody comprises a fleshy context producing many, large spines, up to 40 mm long. The Spine Fungi grow on wood, lack a true stipe and often have only a reduced pileus. All the remaining Tooth-

Fungi grow on the ground, form a pileus and a central to excentric stipe, and the hymenophore is formed on the underside of the cap.

The teeth, or spines, of the hymenophore are of simple construction. Groups of hyphae on the underside of the cap become positively geotropic. These downward, vertically orientated hyphae grow in a strictly longitudinal direction, without any interweaving. Hyphae towards the subhymenial layer are more compact, owing to more frequent branching. The formation of a well-defined subhymenial layer ensures eu-hymenial development. The hymenium consists only of basidia, apart from the Spine Fungi, which are characterized by the inclusion of gloeoplerous elements, usually forming gloeocystidia. Gloeocystidia are recognized by their unlimited growth, often resulting in a series of constrictions in their outline, and the refractive, oily contents. Harrison (1961) defined two types of growth pattern amongst the stipitate hydnoid fungi:

(i) **Determinate Growth** (Fig. 2A): the pileus and stipe of the fruitbody are differentiated at a primordial stage. The margin of the cap is downcurved, often inrolled in the earlier stages, and has a limited capacity to expand, so the cap remains pulvinate or, at the most, depressed in the centre. The hyphal inflation in the context, however, causes rapid expansion, but neighbouring caps are not able to fuse. The context is fleshy and homogeneous, rather brittle, and not zoned. Determinate growth occurs in the Drab and Violet Tooth-Fungi (*Bankera*), the Wood Urchins (*Hydnum*), and the Spiny Cap-Fungi (*Sarcodon*). The Ear Pick-Fungus (*Auriscalpium vulgare* Gray) also exhibits determinate growth but the development of a dimitic hyphal system, with thick-walled skeletal hyphae, results in slower development, and a tougher texture.

(ii) **Indeterminate growth** (Fig. 2B): fruitbody formation is initiated by a vertical hyphal column which eventually expands at the apex to produce the cap. The cap margin grows outwards indefinitely and new teeth continue to be formed on the cap undersurface until the late autumn, when lower temperatures suppress growth. The flesh, or context, which develops slowly, possibly over several months, is tough, and perhaps duplex and zoned. The margin is straight and usually irregular. The stipe continues to thicken and lengthen. Any contact with solid obstacles results in fusion and the formation of irregular, concrescent fruitbodies. Radial and continuous marginal growth results in a pulling apart of the hyphae at the cap centre, so that an irregular, funnel- or vase-shaped fruitbody results, with a disrupted, scaly to scrobiculate surface. Indeterminate growth occurs in the Corky Spine Fungi (*Hydnellum*) and the Scented Spine Fungi (*Phellodon*).

The origins and affinities of the Chanterelles and Tooth-Fungi appear to be diverse. Corner (1957) suggested that a very close relationship existed between *Cantharellus* and the Giant Club (*Clavariadelphus pistillaris* (Fr.) Donk), with the apically truncated, holothecial fruitbody of *Clavariadelphus truncatus* (Quél.) Donk representing an intermediate stage. *Cantharellus* species are more advanced, forming a cap margin (or pileate limb) which is downwardly curved and protecting the inferior hymenium. The stichic basidium of *Cantharellus* is possibly an adaptation, linked to a strongly developed thickening hymenium, rather than a primitive character. *Craterellus*, on the other hand, may be of a different origin, possibly from a series of stereoid, tropical fungi, such as *Cymatoderma* Jungh. and *Podoscypha* Pat. (*Podoscyphaceae*). It is equally possible that both series had a cantharelloid ancestry. There is also the suggestion that the Chanterelles could represent an early stage in the development of the true mushrooms (*Agaricales*). In particular, members of *Hygrophoraceae* are often linked with Chanterelles, with similarities in bright pigmentation, habit, habitat and the elongate basidia. The genera of *Hygrophoraceae*, however, have a fundamental difference in structure, developing a hymenophoral trama, leading to a truly lamellate hymenophore.

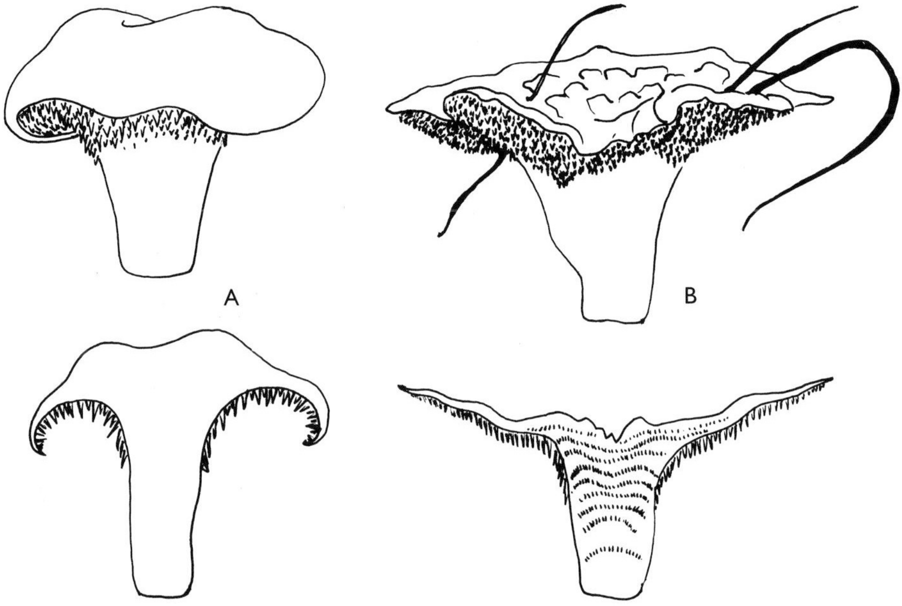

Fig. 2. Growth form in stipitate hydnums. **A,** Determinate growth, with pulvinate cap and downcurved margin; thick, soft-fleshy context lacking zonation; and free of external debris; found in *Bankera*, *Hydnum*, *Sarcodon*. **B,** Indeterminate growth. with depressed, disrupted cap and straight, extending margin; thin, tough context with zonation; external debris incorporated; found in *Hydnellum*, *Phellodon*.

The Wood Urchins (*Hydnaceae*), also have alliances with the Chanterelles, with similarities in the soft-fleshy fruitbodies, inflating hyphae, and large, hyaline, smooth spores. A close association also exists with the family *Scutigeraceae*, and fruitbodies of the poroid genus *Albatrellus* Gray are easily confused with those of *Hydnum*.

The remaining hydnums have quite different affinities, illustrating the convergent evolution of the spinose hymenophore. Most provide intermediate links between genera with a simple, smooth hymenophore and genera with an increasingly complex hymenophore development.

The Spine Fungi (*Hericiaceae* and *Auriscalpiaceae*) possess a unique combination of micro-structures, with hyaline, amyloid spores and a gloeoplerous system. Similar combinations characterize the order *Hericiales*, which includes the *Gloeocystidiellaceae* (resupinate), the *Clavicoronaceae* (clavarioid) and the *Lentinellaceae* (agaricoid). The Corky Spine Fungi (*Hydnellum*) and the Spiny Cap-Fungi (*Sarcodon*) produce brown, gibbous and tuberculate spores, which are the dominating character of the order *Thelephorales*, which includes the genera *Tomentella* Pat. (resupinate), *Lenzitella* Ryvarden (agaricoid) and *Boletopsis* Fayod (poroid). The affinities of the *Bankeraceae* are more problematical. Currently, the family is included within the *Thelephorales*, and similarities between the Scented Spine Fungi (*Phellodon*) and *Sarcodon* cannot be denied. However, the small, hyaline,

echinate spores could link the *Bankeraceae* with *Scytinopogon* Singer in the *Clavariaceae*. Finally, the striking similarity in convergence between the *Hericiales* and the *Thelephorales* should be noted.

Fig. 3. *'Agaricus cantharellus'* from J. Sowerby (1796, tab. 46). 'Very frequent in fir woods, etc. The branching shallow gills, and colour like the yolk of an egg (as Dr. Withering says), will easily distinguish this species. It has a pleasant odour like that of apricots, both when recent and when partly dry, as we observed in some specimens communicated to Lady Arden. This kind of Agaric is frequently eaten, though it often proves rather tough'.

EDIBLE SPECIES

Comparatively easy to recognize, non-poisonous, and often occurring in quantity, Chanterelles and Tooth-Fungi include some of the most sought-after wild fungi in Europe.

By far the favourite is **The Chanterelle**, *Cantharellus cibarius*, one of the few fungi ever to have attracted praise from the notoriously mycophobic British. The English mycologist, M.C. Cooke (1862), noted that 'it is not uncommon to hear from epicures who have been fortunate enough to encounter a well-cooked dish of chantarelles [sic], rapturous encomiums on this golden fungus'; Badham (1847), based on revelations from a banquetting waiter, claimed that in Britain, 'the merits, - nay, the very existence of such a fungus is confined to the Freemasons who keep the secret!'. More prosaically, Berkeley (1860) reported that 'the chantarelle is occasionally served up at public dinners at the principal hotels in London on state occasions, where every effort is made to secure the rarest and most costly dainties'.

Since The Chanterelle is ectomycorrhizal, growing only in association with certain trees, it has never been successfully cultivated, though trials are being made (Dannell, 1994), and hence retains its 'rare and costly' status today. All Chanterelles offered for sale are collected in the wild, and because the fruitbodies are comparatively durable there is a not inconsiderable international trade in the species. Dannell (1994), for example, notes that 60,000 kg of salted Chanterelles were imported into Sweden for canning in 1993, mainly from Poland and Lithuania. Germany imports Chanterelles from North America, and in recent years Scottish Chanterelles have been exported to France, though in case anyone should wish to emulate this, it should be pointed out that collecting wild fungi for reward or sale is illegal in Britain without permission. Fresh (or fairly fresh) Chanterelles are now commonly seen in British delicatessens and specialist greengrocers, albeit at a prohibitive price. The fungus is also sold dried or tinned.

The Chanterelle is said to contain eight kinds of amino acids essential to the human body and, in Chinese medicine, it is claimed that eating it 'frequently prevents abnormal sight opthalmia, night blindness, dryness of skin and mucous membrane from losing its power of secretion' (Ying & al, 1987). The aroma compounds which give The Chanterelle its flavour are complex, but have been studied by Pyysalo (1979).

In the field, the fungus is easy to recognize because of its egg-yellow colour, faint apricot smell, and hymenial ridges which look like gills but are visibly wide, not paper-thin (Fig. 14). The False Chanterelle (Fig. 16) has thin, true gills and is typically bright deep orange, though paler forms are frequent. It is not poisonous, merely unpleasant. However, *Omphalotus illudens* (Schwein.) Sacc., an extremely rare fungus in the south-east of England, is distinctly poisonous and could possibly be mistaken for a Chanterelle. Unlike The Chanterelle, it has true gills and typically grows in clusters at the base of trees and stumps. It is much more common in North America, where it is known as the Jack O'Lantern (the fungus is luminous). A similar species, *Omphalotus olearius* (D.C.: Fr.) Fayod is common with olive trees in southern Europe.

Though all the Chanterelles are edible, only two others are occasionally collected in Europe, namely the **Autumn Chanterelle**, *Cantharellus tubiformis*, and the **Horn of Plenty**, *Craterellus cornucopioides*. The former is a common late season fungus, occurring well into the winter in the south of England, and appreciated mainly for this reason. The latter is local and much less common. It is sometimes collected, but.even in mycophilic parts of Europe is often rejected for its ominous black colour. Cooke (1902) notes that 'no edible fungus is so unattractive as this, which we neglected for years, but at length discovered that we had been deceived by appearances, and had passed over an excellent addition to the table.' Both species occur in North America.

Though some popular American books call it The Sweet Tooth, **The Wood Urchin**, *Hydnum repandum*, has a reputation for being somewhat bitter, at least with age, and as a result is not

quite as highly favoured as the Chanterelles. Nonetheless, Berkeley (1860) wrote that 'few fungi are more delicate than this is when well cooked, the texture and flavour reminding one strongly of a British Truffle, and as it retains its colour, it has a very tempting appearance, which does not prove deceptive on trial'. The Wood Urchin is a substantial fungus and like the Chanterelles the fruitbodies are comparatively long-lasting and free from insects. It is sometimes sold under the French name, 'pied de mouton', though 'sheep's foot' hardly seems an appetizing sobriquet.

Cooke (1862) affirmed 'the esculent qualities' of the **Scaly Tooth**, *Sarcodon imbricatus*, but as an edible species it does not appear to be greatly favoured anywhere, though its potential to lower blood cholesterol levels is being investigated in China (Ying & al, 1987). In Britain, the fungus is distinctly uncommon. None of the remaining Tooth-Fungi are considered edible, being tough, bitter, or variously unpleasant, though not poisonous.

All the **Spine Fungi** are edible at least when young, but are rare and local in Britain and should not be harvested for food. Indeed, the **Monkey Head**, *Hericium erinaceum*, has been proposed for legal protection. However, since Spine Fungi grow on dead or dying wood, they can be cultivated. This is the case in Asia, and more recently the United States, where the Monkey Head is grown on a commercial basis. In China it is considered an effective medicine for ulcers, inflammations, and tumors of the alimentary canal (Stamets, 1993) and can be purchased in pill form, as a powder, as a cordial, or even as a tinned soft drink. As a culinary ingredient, it is perhaps more widely favoured for its attractive appearance than its taste, which can be rather bitter if fruitbodies are not young and fresh.

SAFETY CODE

- **There is no single test for the edibility of wild fungi, many of the 'popular' tests are based upon superstition alone. Each fungus must be accurately identified.**
- **To ensure correct identification, always carefully collect the whole fruitbody, including the base of the stipe. Make a note of the habitat, substratum and any tree association.**
- **Only collect young, fresh fruitbodies for consumption; always ignore old, decaying and mouldy specimens. Remember different species may grow in close proximity and careless collecting could result in a mixture of species.**
- **Never collect near roads or industrial areas as fungi can readily absorb heavy metals, such as lead.**
- **Transport your collections home in a well aerated container, such as a wickerwork basket; avoid placing specimens in polythene bags or enclosed containers, and keep the species separate. Always wash your hands after handling fungi.**
- **Do not eat raw fungi. Never eat a known edible species if it has been mixed with poisonous species in your collecting basket.**
- **If trying a wild fungus for the first time then (a) only take a small sample, (b) do not mix with other species, and (c) retain one fruitbody in case poisoning results and identification needs to be confirmed.**
- **Seek medical help immediately when poisoning is suspected.**

ECOLOGY, DISTRIBUTION AND CONSERVATION

Ecological requirements of the Chanterelles and Tooth-Fungi vary according to species, but most are woodland fungi. Their fruitbodies usually occur on the soil amongst leaf litter, or in some cases (*Hericium* & *Mucronella* spp.) are lignicolous in their development. They are usually developed during the autumn, August–November. However, in one species, *Auriscalpium vulgare*, fruitbodies develop from fallen, often buried, cones of conifers, commonly *Pinus sylvestris*, and may occur almost throughout the year. Amongst the Tooth-Fungi, species of *Hydnum*, *Hydnellum*, *Phellodon*, *Sarcodon* and *Bankera* are considered to be ectomycorrhizal, although this is based largely on field observation and is proven only for *Bankera fuligineoalba* (Arnolds, 1989). Species generally occur in association with particular trees and especially, in many cases, are inhabitants of ancient or at least mature (> 40 years old) forest systems. Soil type is also of ecological importance to many of these fungi. For example, *Phellodon* species are generally associated with poor, acid soils, whereas a wider ecological range is found within the genera *Hydnellum*, *Sarcodon* and *Hydnum*, where some species are characteristic of mineral-rich soils.

Amongst the *Cantharellales* other than *Hydnum*, *Cantharellus* species are also ectomycorrhizal. The Chanterelle (*C. cibarius*) has a broad host range but optimally requires well-drained, nutrient-rich sandy soils with a pH of about 4.5 (Jansen & van Dobben 1987). Other species of this group are generally more host restricted and may favour either calcareous or acidic soils. *Craterellaceae* are non-mycorrhizal.

Worldwide, the genera of Chanterelles and Tooth-Fungi are widely distributed especially in temperate regions, and many species are common to most of Europe and to North America. *Cantharellus*, *Craterellus*, *Pseudocraterellus*, *Auriscalpium* and *Mucronella* are cosmopolitan in distribution, the first in particular being well speciated throughout temperate and tropical areas and indeed reaching greatest speciation in tropical Africa. *Hydnum*, *Phellodon* and *Sarcodon* are represented in the southern hemisphere by a few species. In the tropics, Tooth-Fungi of these genera as well as *Hydnellum* are comparatively rare, whilst the small genera *Bankera* and *Hericium* are known only from north temperate areas. Within the British Isles, individual species exhibit a range of distribution patterns according to their ecological requirements as outlined above. Whilst some of the Chanterelles, for example, are widely distributed though sometimes rare throughout the British Isles, a number of hydnaceous species are restricted to native pine forest and are found only in the Scottish Highlands. Species of *Hericium* are uncommon in England, occurring mostly in the south, and are not recorded from Scotland. They appear to be at the edge of their range, possibly restricted by climatic conditions.

A marked decline in the appearance of many species of *Cantharellaceae* and hydnaceous fungi, especially of those presumed to form mycorrhizal associations, has been noted in many parts of Europe during the past 20–30 years. This decline has been well documented for the stipitate hydnaceous species in the Netherlands (Arnolds, 1989). Many of these fungi now appear in various European red data lists as endangered or vulnerable, and 22 of the British species are considered as either extinct or threatened with extinction in some part of their European range (Lizon, 1995). In Denmark, for example, both *Sarcodon glaucopus* and *S. scabrosus* are considered extinct (Vesterholt & Knudsen, 1990), and in the Netherlands, five species of *Hydnellum* as well as *Bankera fuligineoalba*, *Phellodon tomentosus* and *Pseudocraterellus cinereus* are also listed as extinct. In the British Isles a similar situation exists; numerous species appear to be increasingly scarce and there is a lack of recent records for many of them. This is reflected in the listing of 23 species, mostly in the endangered or vulnerable categories, in the Provisional British Red Data List (Ing, 1992). The true situation

in the British Isles is difficult to assess in the absence of effective experimental data. Many species are rare and fruiting is known to be strongly influenced by weather conditions (Arnolds, 1989); abnormally hot and dry seasons have become more frequent and may prevent fruiting completely, particularly for mycorrhizal species. However, it is well established that many of these fungi, notably the stipitate hydnaceous species and The Chanterelle, are adversely affected by pollution, particularly by increased acidification and eutrophication of the soil. There is cause for concern and further research into these, and indeed the perceived decline of macrofungi in general, is desirable.

Decline in the abundance of *Cantharellus cibarius* has also been indicated, and it is likely that this is due to similar problems. In the Netherlands the decline of this species was studied by Jansen & van Dobben (1987) who found that, although changes in forest management practice and natural succession were factors in particular sites, soil acidification was the major factor influencing decline. These authors also showed that neither overcollecting nor the build-up of heavy metals were significant factors in this case. The decline of this species in Europe is further discussed by Danell (1994), who noted that increased nitrogen deposition also has adverse affects on *C. cibarius*, and that habitat destruction may also be an important factor in the decline of this and other mycorrhizal species. Overcollecting of edible fungi, with consequent site trampling, have been considered as likely factors having adverse effects on many macrofungi.

Conservation measures for protection of Chanterelles and Tooth Fungi may need to be considered, although legal protection alone or site protection and management in the absence of appropriate control of pollutants are considered insufficient (Arnolds, 1989). Nevertheless, legal protection has now been implemented for some species: for example, *Sarcodon imbricatus* is included in the list of fungi which appear on Appendix 1 of the *Berne Convention on the Conservation of European Wildlife*, and *Hericium erinaceum* is amongst the twelve fungi proposed for protection under the latest review of the *Wildlife & Countryside Act.*

COLLECTING CODE

- **Consider the landowner; seek permission to collect fungi. When walking over farmland, always keep to marked foot-paths. Do not leave litter, and avoid the risk of fire.**
- **Always use gates and styles provided; make sure that gates are properly closed behind you.**
- **Observe the local bylaws. Trespass is a civil offence in England and Wales.**
- **The intentional uprooting of any plant without permission is prohibited (The *Wildlife & Countryside Act*, 1981). 'Uprooting' with regard to fungi could clearly be contentious. Many sites have their own local bylaws, which may prohibit the collection of fungi (or forest produce).**
- **'A person who picks mushrooms growing wild on any land..., does not steal what he picks unless he does it for reward or for sale or for other commercial purpose...For the purpose of this section 'mushroom' includes any fungus' (Theft Act, 1968).**
- **Do not overcollect; only gather as much as you need.**

COLLECTING, EXAMINATION AND PRESERVATION

As with all fungi, it is important to understand the range of habitats favoured by the various Chanterelles and Tooth Fungi if particular species are being sought. The ecological requirements of the various species are outlined in the previous chapter and more fully discussed under each entry. To obtain fresh material, most species should be sought in the autumn months, although *Auriscalpium vulgare* may occur almost throughout the year. When collecting, remember always to seek the landowners permission in advance. Unauthorized collecting for commercial purposes is illegal in Britain.

In all cases, only material in good, fresh condition should be collected. This is obviously important for Chanterelles and other fungi being collected for the table, but applies equally to specimens for herbarium preservation. The fruitbodies of most of the species are fleshy and will break down and rot fairly rapidly after maturing. Although some may remain until the following spring, old specimens are of little value and should be avoided. They will have lost most of their important characters so that accurate identification is often impossible. Many of the Tooth-Fungi when actively growing offer valuable characters such as colour and colour changes of the flesh, odour and even taste which are essential for accurate specific identification and which cannot be obtained from old or dried specimens. Material should be collected with care. Whole fruitbodies should be obtained, but for terricolous species these should be collected with minimum disturbance to the soil so as not to damage the growing mycelium. The host wood for lignicolous species should be ascertained wherever possible. Fruitbodies should be collected into individual containers or a flat-bottomed basket, taking care to keep different collections separate. Avoid polythene bags which cause sweating and rapid decay of the specimens. Notes as to habitat and substrate are important for all species collected. From fresh material notes should be made as to size range, cap and stipe colours and surface characters and, for Tooth-Fungi and Spine-Fungi, as indicated above, colour and colour changes in the flesh and on drying.

For herbarium preservation, specimens should be dried as rapidly as possible and then deep frozen for at least 48 hours to rid them of possible pests, and should be maintained in dry conditions. Each specimen should be housed in a packet or box, and clearly labelled with full details as to habitat, locality, date, and collector, together with any relevant descriptive notes.

Identification of specimens in most cases requires, in addition to the field characters noted, the use of light microscopy for examination of spores and hyphal characters. For fresh specimens, examination in distilled water is recommended, but KOH should be used for dried material in order to fully rehydrate and swell the hyphal elements. In the case of the Spine Fungi (*Hydnellum* and *Phellodon* spp.), colour reaction of the context with alkali solution is an important characteristic in some species, notably *H. concrescens* and *P. niger* amongst the British ones. A dark greenish stain is rapidly exuded from these species when tissue from the spines or context is placed in KOH or ammonia solution. Stain reactions are in general of little importance in the examination of these fungi. However, in the *Hericiales*, spores and hyphae of some species are amyloid so that use of Melzer's Reagent for examination of these is essential.

KEYS TO ORDERS AND FAMILIES

(British hydnoid representatives)

1. Spores hyaline, smooth, inamyloid; fruitbody stipitate, mushroom-like to funnel-shaped, terrestrial and ectomycorrhizal; growth determinate, with down-curved cap margin; context soft-fleshy to membranaceous; hymenophore smooth, veined, 'falsely lamellate' or toothed .... - **Chanterelles** & **Wood Urchins** (*Cantharellales*, p. 14) **2**
1. Spores either amyloid or tuberculate; fruitbody effused to stipitate, terrestrial and ectomycorrhizal or lignicolous; growth determinate or indeterminate; eu-hymenial, not thickening; hymenophore toothed **4**

2. Hymenium thickening, not eu-hymenial; hymenophore smooth, wrinkled or 'falsely-lamellate' **3**
2. Hymenium not thickening, eu-hymenial; hymenophore toothed; fruitbody fleshy, mushroom-like .... - **Wood Urchins** (*Hydnaceae*) p. 39

3. Clamp-connexions present; hymenophore radially veined to 'falsely lamellate'; fruitbody mushroom-like to funnel-shaped, often with bright pigments .... - **Chanterelles** (*Cantharellaceae*) p. 22
3. Clamp-connexions absent; hymenophore smooth, rugulose or 'falsely lamellate'; fruitbody tubular to funnel-shaped, dull with grey or brownish tints .... - **Horn of Plenty** (*Craterellaceae*) p. 15

4. Fruitbody lignicolous; spores amyloid, hyaline; refractive gloeoplerous elements present; clamp-connexions present.................. - **Spine Fungi** (*Hericiales* p.p., p. 44) **5**
4. Fruitbody terrestrial or in litter; spores inamyloid, yellowish brown or hyaline, echinulate, verrucose or tuberculate; gloeoplerous elements absent; hymenophore smooth, papillate or toothed; clamp-connexions present or absent .... - **Tooth-Fungi** & **Spiny Cap-Fungi** (*Thelephorales* p.p., p. 58) **6**

5. Fruitbody either bracket-like or of isolated, pendant spines; context soft-fleshed, with monomitic hyphal system; generative-hyphae inflating; on dead wood .... - **Spine Fungi** (*Hericiaceae*) p. 48
5. Fruitbody pileate with a long, thin stipe; context tough, with a dimitic hyphal system, with skeletal hyphae; generative-hyphae not inflating; on fallen conifer cones .... - **Ear Pick-Fungus** (*Auriscalpiaceae*) p. 45

6. Spore deposit whitish; spores hyaline, echinulate; clamp-connexions absent; fruitbody with odour of fenugreek when dried......... - **Scented Spine Fungi** (*Bankeraceae*) p. 59
6. Spore deposit brown; spores brown, tuberculate; clamp-connexions present or absent .... - **Corky Spine Fungi** & **Spiny Cap-Fungi** (*Thelephoraceae*) p. 74

CHANTERELLES and WOOD URCHINS

CANTHARELLALES Gäum.

Vergl. Morph. Pilze: 495 (1926).

Fruitbody pilothecial, agaricoid forms pileate with a central or excentric stipe. *Hymenophore* smooth, rugose or radially ridged, simulating 'false-lamellae', with an obtusely rounded, fertile edge. *Context* homoiomerous, fleshy to membranous; *hyphal system* monomitic, rarely dimitic, with or without clamp-connexions. *Hymenium* thickening. *Spores* ballistosporic, subglobose to elongate ellipsoid, hyaline to pale ochraceous, inamyloid, non-dextrinoid. *Basidia* slender, elongate (50–100 µm), clavate-cylindrical, bearing 2–8 sterigmata; stichic (with longitudinal meiotic spindle). *Cystidia* not developed. *Development* gymnocarpic, epigeous. Terricolous, humicolous, never lignicolous. World-wide. *Type family: Cantharellaceae* Herter.

Generally, the fruitbodies are short-lived, with a soft and fleshy consistency, composed of thin-walled, inflating, generative-hyphae. No skeletal- or ligative-hyphae are developed. The two, related, distinguishing characteristics for the order are the very elongated basidia and the thickening hymenium. This structural combination separates the *Cantharellales* from genera of the mushrooms or *Agaricales*, to which they have been often linked in the past. The inflating generative-hyphae distinguish the order from the other aphyllophoroid orders.

This small group contains some of the most highly prized of the edible fungi. Owing to their inability to utilize cellulose and similar complex molecules, the fruitbodies cannot be readily cultivated, but depend on forming ectomycorrhizal relationships with the roots of forest trees. The fruitbodies, however, are gathered on a commercial basis in many parts of the world, especially in northern Europe and North America. There are around 100 species world-wide and greatest speciation is found in equatorial Africa, especially in the open, *Brachystegia-Isoberlinia-Julbernardia* woodland, known as 'miombo', which runs from southern Tanzania to the Republic of South Africa. The bright yellow, red, orange and pink pigments of the African species have made them increasingly popular in European restaurants.

Key to Families

1. Hymenium thickening; hymenophore smooth, wrinkled or falsely-lamellate .... - 2
1. Hymenium not thickening; hymenophore toothed .... 3. **Wood Urchins** (*Hydnaceae*) p. 39

  2. Clamp-connexions present; hymenophore radially veined to falsely-lamellate; fruitbody mushroom-like to funnel-shaped, often with bright pigments
  .... 2. **Chanterelles** (*Cantharellaceae*) p. 22
  2. Clamp-connexions absent; hymenophore smooth, rugulose or falsely-lamellate; fruitbody tubular to funnel-shaped, dull with grey or brownish tints
  .... 1. **Horn of Plenty** (*Craterellaceae*) p. 15

1. CRATERELLACEAE Herter

Kryptog. - Fl. Brandenb. 6: 141 (1910).

Fruitbody solitary or caespitose; funnel-shaped, with or without a tubular stipe. *Hymenophore* pale to dark coloured, smooth or rugulose. *Context* membranous to subcoriaceous; *hyphal system* monomitic; *generative-hyphae* more or less inflated, many developing a thickened wall, sometimes abundantly secondarily septate, lacking clamp-connexions. *Hymenium* thickening, at least to some degree. *Spores* subglobose to ellipsoid, hyaline, inamyloid, non-dextrinoid, thin-walled, smooth. *Basidia* cylindrical to narrowly clavate, bearing 2–6, often curved, sterigmata. *Cystidia* absent. Terricolous or humicolous. *Type genus: Craterellus* Pers.

Key to Genera

1. Pileus tubular from the first, later becoming pileate through marginal expansion, lacking a true stipe; hyphae lacking conspicuous secondary septation .... **1. Craterellus**

1. Initially pileate and with a distinct, slender, solid stipe; pileus becoming depressed and funnel-shaped, often perforate with the stipe; hyphae with abundant secondary septa .... **2. Pseudocraterellus**

1. CRATERELLUS Pers.,

Mycol. Eur. 2: 4 (1825).

Trombetta Kuntze, Rev. Gen. Pl. 1: 873 (1891).
Cantharellus Fr. subg. *Craterellus* (Pers.) Kühner & Romagn., *Fl. Anal. Champ. Sup.*: 47 (1953).

Fruitbody often caespitose, funnel-shaped with a basal tubular stipe; upper (inner) surface sterile, smooth to finely squamulose; margin straight. *Hymenophore* on lower (outer) surface, smooth or somewhat rugulose, owing to uneven thickening of underlying context. *Context* thinly fleshy to subcoriaceous; *hyphal-system* monomitic; *generative-hyphae* more or less inflating, often with thickened walls; clamp-connexions absent. *Spore-deposit* white to pale cream. *Spores* subglobose, ovoid to oblong-ellipsoid, hyaline, inamyloid, thin-walled, smooth. *Basidia* elongate, subcylindrical, bearing 2–6 curved sterigmata. *Hymenium* slightly thickening. *Cystidia* absent. Terricolous, cosmopolitan. *Type species: C. cornucopioides* Pers.

HORN OF PLENTY

Craterellus cornucopioides (L.: Fr.) Pers., *Mycol. Eur.* 2: 5 (1825).
Cantharellus cornucopioides L.: Fr., *Syst. Mycol.* 1: 321 (1821); L., *Fl. Dan.*: pl. 1181 (1753, as *Peziza*).

Pileus 2–8 cm diam., 8–10(–20) cm tall, deeply funnel- to trumpet-shaped; surface black to blackish brown, drying greyish brown, rugulose-scurfy to felty-tomentose, often with a curled, undulating, down-curved margin. *Hymenophore* smooth to slightly veined-wrinkled, sometimes with faint anastomosing ridges, pale brown to grey-brown. *Stipe* not truly developed but formed by the extending pileus, tubular, tapering below; total height up to 12 cm. *Context* thin, up to 1 mm thick, soft-leathery, grey; odour pleasant and fruity; *taste* pleasant; *generative-hyphae* 2–7 µm diam., mostly not inflating or inflating slightly up to 10 mm diam., hyaline, thin-walled, lacking clamp-connexions. *Spore deposit* off-white to pale cream colour. *Spores* 10–16(–17) × 6–9(–11) µm, ovoid to broadly ellipsoid, hyaline, thin-walled, smooth, filled with small to large, refractive guttules. *Basidia* 70–95 × 8–10 µm, elongate clavate, generally bisporic. *Subhymenial layer* of loosely woven hyphae, 2–5 µm diam. *Pileipellis* a disrupted epicutis of radially parallel hyphae, formed of slightly inflated elements, 25–50 × 5–15 µm, with a slightly thickened, brown wall, lacking clamp-connexions.

Habitat: gregarious, occurring in both deciduous and coniferous woodland, but especially common in woodlands of hazel (*Corylus*), beech (*Fagus*), and oak (*Quercus*). It also grows in mossy pine woods.

Distribution and frequency: widespread throughout Europe and North America, although *Craterellus fallax* A. H. Smith is the more common species in the eastern states of the U.S.A. August to November.

Other remarks: good, edible species. although the dark colours have given it the names of 'Trompe des Mortes' and 'Black Trumpet'. Known in France as 'la viande des pauvres' ('poor man's meat'), it can be dried and preserved for subsequent use in stews. Not to be confused with the much rarer Grey Chanterelle (*Pseudocraterellus cinereus* (Pers.: Fr.) Kalamees) which has a pseudolamellate hymenophore.

Good colour illustrations may be found in Breitenbach & Kränzlin (1986: pl. 487); Comitato di Studio del Gruppo di Vicenza (1989: 222); Nilsson & Persson (1978: 68–69); Persson & Mossberg (1994: 68).

Fig. 4. *Craterellus cornucopioides* (England, Is. Wight, Firestone Copse, 3 Nov. 1984, B.M. Spooner).

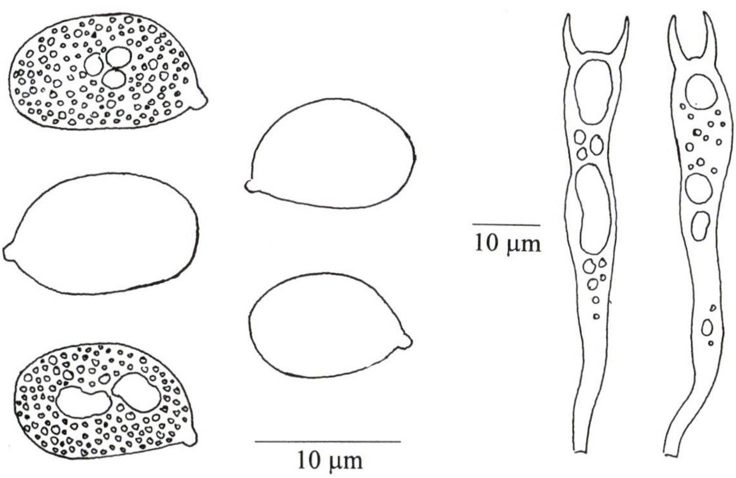

Fig. 5. *Craterellus cornucopioides* (England, Hampshire, New Forest, 14 Oct. 1995, N.W. Legon, K(M) 31628) spores; bisporic basidia.

2. PSEUDOCRATERELLUS Corner

in *Sydowia* ser. 2, Beih. 1: 268 (1957).

Pileus soon funnel-shaped and often pervious with stipe, at times spathulate to flabelliform; margin soon straight. *Hymenophore* smooth or faintly rugulose. *Context* subcoriaceous, not hygrophanous; *hyphal system* monomitic; *generative-hyphae* thin-walled, with short elements, often secondarily septate; clamp-connexions absent. *Spore deposit* white or yellowish. *Spores* ellipsoid, hyaline, inamyloid, thin-walled, smooth, multiguttulate. *Basidia* elongate, subventricose, bearing 2–4 (rarely 5–6) sterigmata. *Hymenium* usually thickening. *Cystidia* none. Cosmopolitan but rare. *Type species*: *P. sinuosus* (Fr.) D.A. Reid.

Key to Species

1. Entire fruitbody grey-brown to blackish; pileus 2–4(–6) cm diam., always funnel-shaped; hymenophore strongly ridged, anastomosing, ash-grey to bluish grey; stipe grey to blackish; spores 7.5–12.5 × 5–7 µm .... **1. Grey Chanterelle** (*P. cinereus*)
1. Pileus greyish to sepia brown, 1–5 cm diam., funnel-shaped, at times perforate; hymenophore wrinkled and interveined, greyish ochre; stipe yellow to pale brown; spores 8.5–13 × 6–9 µm .... **2. Wavy-capped Chanterelle** (*P. sinuosus*)

1. GREY CHANTERELLE

Pseudocraterellus cinereus (Pers.: Fr.) Kalamees in *Tartu rük. ülok. Toim.* 136, bot. Works, 6: 90 (1963).
Merulius cinereus Pers., *Syn. Meth. Fung.*: 490 (1801).
Helvella hydrolips Bull., *Herb. Fr.*: pl. 465/2 (1789).
Cantharellus cinereus Pers.: Fr., *Syst. Mycol.* 1: 320 (1821).
Craterellus cinereus (Pers.: Fr.) Quél., *Fl. Mycol. Fr.*: 36 (1888).

Pileus 2–4(–6) cm diam., convex then expanded; surface grey-brown to blackish brown, perforate at the centre, radially rugose or striate, smooth towards the margin; margin undulate. *Hymenophore* decurrent, slightly radially ridged to almost pseudolamellate, about 1 mm thick, with occasional transverse interveining, grey to bluish grey. *Stipe* 3–6 × 0.5–0.8 cm, tubular, tapering below, hollow; surface grey-brown then blackish, fibrillose, veined-rugose. *Context* thin, grey-brown to blackish; *odour* fruity, *taste* mild; *generative-hyphae* 2–14 µm diam., hyaline, thin-walled, lacking clamp-connexions. *Spore deposit* off-white. *Spores* 7.5–10.5(–12.5) × 5–7 µm, ellipsoid to oblong-ellipsoid, hyaline, thin-walled, with strongly refractive guttules. *Basidia* 50–65 × 7–10 µm, elongate clavate, bearing 4, 5 or 6 sterigmata. *Subhymenial layer* loosely woven. *Pileipellis* a poorly defined cutis of agglutinated, radially parallel hyphae, 3–9 µm diam., with thin, brown walls, lacking clamp-connexions.

Habitat: clustered or solitary, on soil and amongst moss, generally in beech (*Fagus*) woods.

Distribution and frequency: found throughout Europe and northern North America but never abundant. Rare to occasional throughout the British Isles.

Other remarks: *Craterellus cornucopioides* grows in association with this species but is generally more robust, has a smooth hymenophore, and lacks a true stipe.

Good colour illustrations may be found in Breitenbach & Kränzlin (1986: pl. 486); Cetto (1987a: 128), Comitato di Studio del Gruppo di Vicenza (1989: 221); Persson & Mossberg (1994: 64).

Fig. 6. *Pseudocraterellus cinereus* (England, Essex, near Colchester, Sep. 1984, B.M. Spooner).

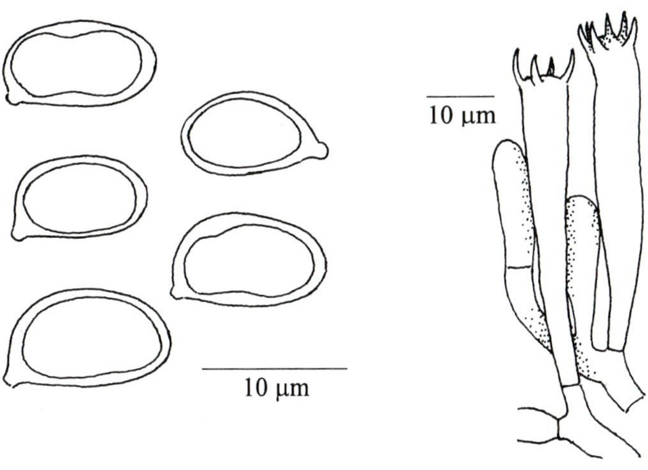

Fig. 7. *Pseudocraterellus cinereus* (England, Gloucestershire, Cirencester Park, 21 Nov. 1982, G. Stone, K(M) 34140) spores; basidia.

2. WAVY-CAPPED CHANTERELLE

Pseudocraterellus sinuosus (Fr.) D.A. Reid in *Persoonia* 2: 122 (1962); Corner ex Heinem. in *Bull. Jard. Bot. Etat. Brux.* 28: 426 (1958, nom. nud.).
Cantharellus sinuosus Fr., *Syst. Mycol.* 1: 319 (1821).
Cantharellus pusillus Fr., *Syst. Mycol.* 1: 319 (1821).
Merulius undulatus Pers., *Mycol. Eur.* 2: 20 (1825).
M. tubiformis var. *crispus* Bull. ex L. March. in *Bijdr. natuurk. Wetensch.* 3: 272 (1828).
Craterellus sinuosus (Fr.) Fr., *Epicrisis*: 533 (1838).
Cantharellus crispus (Bull. ex L. March.) Fr., *Epicrisis*: 533 (1838).
Craterellus crispus (Bull. ex L. March.) Berk., *Outl. Brit. Fung.*: 266 (1860).
Craterellus floccosus Quél. in *Bull. Soc. Bot. Fr.* 24: 308 (1877).
Craterellus sinuosus var. *crispus* (Fr.) Quél., *Fl. Mycol.*: 35 (1888).
Pseudocraterellus undulatus (Pers.) Rausch var. *sinuosus* (Fr.) Bon in *Doc. Mycol.* 22 (fasc. 88): 37 (1993).

Pileus 1–5 cm diam., thin-fleshed, funnel-shaped; surface greyish brown to sepia brown, becoming paler to whitish, at times perforate; margin undulate and crisped. *Hymenophore* decurrent, wrinkled and somewhat interveined, greyish ochre. *Stipe* 3–6 × 0.3–0.8 cm, cylindrical, tapering below, pitted, hollow; surface yellowish to pale brown. *Context* thin, soft-fibrous; *odour* pleasant; *taste* mild; *generative-hyphae* 3.5–13 µm diam., broad but not significantly inflated, with short segments and abundant septa, including secondary septa, lacking clamp-connexions. *Spore deposit* white. *Spores* 8.5–13 × 6–9 µm, short ovoid to ellipsoid, hyaline, thin-walled, smooth, with large refractive guttules. *Basidia* 50–80 × 6–11 µm, elongate clavate, mostly bearing 2 or 4, occasionally 6, sterigmata. *Subhymenial layer* loosely woven. *Pileipellis* an indefinite cutis of interwoven hyphae, 4–9 µm diam., similar to those of context but with brown, granular encrustations.

Habitat: solitary or more usually clustered, occasionally in groups of 100 or more, on soil amongst leaf-litter in deciduous woods, particularly associated with hazel (*Corylus*), beech (*Fagus*), ash (*Fraxinus*), and oak (*Quercus*).

Distribution and frequency: occurs throughout northern and central Europe; uncommon to locally abundant throughout the British Isles. Only rarely recorded from North America.

Other remarks: similar in appearance to *Cantharellus aurora* (Batsch) Kuyp. and *C. tubiformis* Bull.: Fr., differing macroscopically in the generally shorter and thinner stipe, the very wavy pileal margin, and often clustered habit.

Good colour illustrations may be found in Breitenbach & Kränzlin (1986: pl. 488); Cetto (1987a: 138), Comitato di Studio del Gruppo di Vicenza (1989: 223).

Fig. 8. *Pseudocraterellus sinuosus* (England, Surrey, Norbury Park, N.W. Legon).

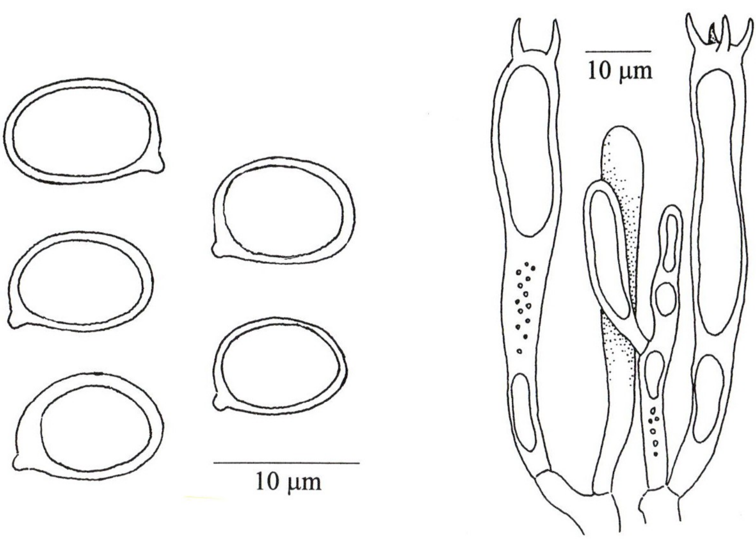

Fig. 9. *Pseudocraterellus sinuosus* (England, Sussex, Fernhurst, 2 Sep. 1968, D.A. Reid, K(M) 33980) spores; basidia.

2. CANTHARELLACEAE J. Schröt.

in Cohn, *Kryptog.-Fl. Schles.* 3(1): 507 (1888).

Fruitbody agaricoid, tubular to funnel-shaped or pileate with a sterile stipe, pale or brightly coloured. *Pileus* membranous to fleshy-coriaceous, putrescent, with sterile upper surface. *Hymenophore* smooth becoming rugose or folded to form radial 'false-lamellae' or ridges, owing to the thickening hymenophore, often interveined, with an obtuse, fertile edge. *Veils* absent. *Context* monomitic; *generative-hyphae* inflating, with clamp-connexions. *Spore deposit* white, pale ochraceous or salmon-pink. *Basidiospores* subglobose to ellipsoid, hyaline, inamyloid, non-dextrinoid, smooth, thin-walled. *Basidia* narrowly clavate, bearing 2–8 sterigmata. Gymnocarpic. Terricolous or humicolous. World-wide. *Type genus: Cantharellus* Juss.: Fr.

CANTHARELLUS Juss.: Fr.

Syst. Mycol. 1: 316 (1821); Juss, *Gen. Pl.*: 4 (1789).

Pileus soft fleshy, often funnel-shaped with an incurved margin; often brightly coloured, white, yellow, orange or red. *Hymenophore* decurrent, smooth or forming radial ridges, with a thick, fertile edge. *Stipe* meso- to pleuropodal. *Context* putrescent, with a monomitic hyphal system of inflated, thin-walled generative-hyphae; clamp-connexions present. *Hymenium* thickening. *Basidia* very elongate clavate, bearing 2–8 sterigmata, and a basal clamp-connexion. *Cystidia* absent. *Subhymenial layer* loosely woven and indefinite. *Pileipellis* a repent cutis or trichodermium. World-wide. *Type species: C. cibarius* Fr.

Corner (1966) divided the genus into two subgenera, based upon pigmentation. Brightly coloured species, with yellow, orange, pink, red or white pigmentation were placed in subgenus *Cantharellus*, and included *C. amethysteus*, *C. cibarius*, *C. friesii*, and *C. pallens*. These tend to be thick, fleshy species, with a solid stipe. Species with brown, grey, fuscous or fuliginous pigments were placed in a new subgenus, *Phaeocantharellus* Corner, and these included *C. aurora*, *C. melanoxeros* and *C. tubiformis*. These are thin-fleshed species with hollow stipe and often with a perforate pileus. Peck (1887), however, had already proposed the subgeneric name *Leptocantharellus* for the latter group. *Leptocantharellus* was also chosen by Donk (1969), who suggested a further practical, rather than taxonomic, division of the subgenus. He placed those species with a 'falsely lamellate' hymenophore into the Stirps *Lepto-Plicati* e.g. *C. tubiformis*, *C. melanoxeros*, and species with a vein-like hymenophore were placed into Stirps *Lepto-Phlebini* e.g. *C. aurora*. Fiasson (1973) presented a biochemical distinction between the subgenus *Cantharellus*, with bicyclic carotenes, and the subgenus *Leptocantharellus* (as *Phaeocantharellus*), with aliphatic carotenes.

Although the number of species in Britain and northern Europe is relatively few, this is not the case elsewhere. In equatorial Africa, 35 species have been described. They are particularly abundant in Malawi, southern Tanzania, Zambia and Zimbabwe, where many of the brightly coloured species, such as *C. symoensii* Heinem. (orange-red), *C. longisporus* Heinem. (pinkish orange) and *C. miniatescens* Heinem. (reddish orange) are collected on a commercial scale for west European restaurants.

Key to Species

1. Fruitbody robust, fleshy; pileus often 10 cm diam. or more; stipe usually solid; hymenophore strongly plicate to ridged, yellow to egg-yellow **2**
1. Fruitbody not fleshy; pileus usually funnel-shaped, rarely up to 10 cm diam.; stipe hollow **4**

2. Pileus white to pale yellow, sometimes with a pinkish tint, quickly bruising rusty brown, up to 15 cm diam., stipe stout .... **1. Pale Chanterelle** (*C. pallens*)
2. Pileus and stipe pure yellow to egg-yellow, not strongly bruising **3**

3. Pileus with a disrupted, pinkish-lilac, violaceous to mauve covering layer; uncommon .... **2. Amethyst Chanterelle** (*C. amethysteus*)
3. Pileus lacking any pinkish or violaceous tints; common **3. The Chanterelle** (*C. cibarius*)

4. Context and stipe surface finally blackening; hymenophore pinkish yellow; pileus 2–6(–10) cm diam., pale saffron yellow **4. Blackening Chanterelle** (*C. melanoxeros*)
4. Context not blackening **5**

5. Hymenophore smooth, finally weakly veined, flesh-pink to salmon pink; pileus 2–6 cm diam., dark brown, funnel-shaped, sometimes perforate; stipe golden-yellow; rare, conifer-wood species .... **5. Yellow-stemmed Chanterelle** (*C. aurora*)
5. Hymenophore veined to plicate, even when young, often furcate **6**

6. Hymenophore strongly plicate, pinkish orange becoming ochraceous; pileus 1–4.5 cm diam. at the most, orange to pinkish orange, umbilicate, membranous-fleshy; stipe concolorous with pileus, whitish at base; solitary or in small groups .... **6. Orange Chanterelle** (*C. friesii*)
6. Hymenophore veined to plicate, dull yellowish becoming grey; pileus usually perforate; stipe yellow, greyish brown above, sometimes entirely grey **7**

7. Stipe greyish yellow to orange; pileus 1.5–6 cm diam., grey-brown to rusty brown, fibrillose scaly; spores 9–12.5 × 7–10 μm .... **7a. Autumn Chanterelle** (*C. tubiformis* var. *tubiformis*)
7. Stipe pale yellow; pileus pale yellow to greyish brown .... **7 b.** *C. tubiformis* var. *lutescens*

1. PALE CHANTERELLE

Cantharellus pallens Pilàt in *Omag.Traian Săvulesc*: 600 (1959).
C. ferruginascens P.D. Orton in *Notes Roy. Bot. Gard., Edinb.* 29: 83 (1969).
C. cibarius Fr. var. *ferruginascens* (P.D. Orton) Courtec. in *Doc. Mycol.* 23 (fasc. 91): 3 (1993).

Pileus 2–7(–15 cm) diam., convex becoming depressed to umbilicate or funnel-shaped on expansion, rarely subumbonate; surface ochraceous buff to pale lemon yellow or olive-beige, sometimes with a pinkish tint, paler on drying, discolouring rusty ochraceous to bright golden brown on bruising and drying, dry, silky smooth to slightly floccose at the centre; margin whitish, thin, irregularly lobed and wavy. *Hymenophore* deeply decurrent, narrow, plicate to ridged, often furcate towards margin and interveined, fairly densely crowded, pale yellowish buff to pale lemon-yellow, bruising apricot yellow then rusty ochraceous. *Stipe* 2–4 × 0.4–1.5 cm, cylindrical or tapering below, firm, solid; surface whitish or concolorous with pileus and bruising rusty ochraceous, smooth to subfloccose. *Context* up to 1.0 cm thick, firm fleshy, cream to yellowish cream; *generative-hyphae* 3–13 µm diam., hyaline, thin-walled, somewhat inflated to appear constricted at septa, with prominent clamp-connexions; *odour* indistinct, pleasant; *taste* mild or slightly bitter. *Spore deposit* pale cream to yellowish cream. *Spores* 7.5–10.5 × 4.5–6.5 µm, broadly ellipsoid to oblong ellipsoid, hyaline, thin-walled, with refractive contents. *Basidia* 55–80 × 7–10 µm, elongate clavate, bearing 5–6 sterigmata; with conspicuous basal clamp-connexion. *Subhymenial layer* broad, indistinct, loosely woven. *Pileipellis* a cutis of interwoven hyphae, 3–8 µm diam., hyaline or pale yellowish, with a thin to partially thickened wall (–1.0 µm), with clamp-connexions.

Habitat: on calcareous soil, in mixed, deciduous woodland.

Distribution and frequency: rare, northern European species, appearing at beginning of July through to September. British records confined to southern England.

Other remarks: more robust than The Chanterelle (*C. cibarius*) and much paler, especially the hymenophore. The reddish brown to golden brown bruising is a distinctive feature of this species. *Cantharellus cibarius* Fr. var. *alborufescens* Malenç., a Mediterranean variety, described from Morocco, is very similar but whiter with rufous spots, and has more cylindrical spores, 9–12 µm long.

Good colour illustrations may be found in Courtecuisse (1993: pl. 3/2); Nilsson & Persson (1978: 70–71); Persson & Mossberg (1994: 40) and Phillips (1981: 190).

Fig. 10. *Cantharellus pallens* (England, Devon, Trendlebere, N.W. Legon).

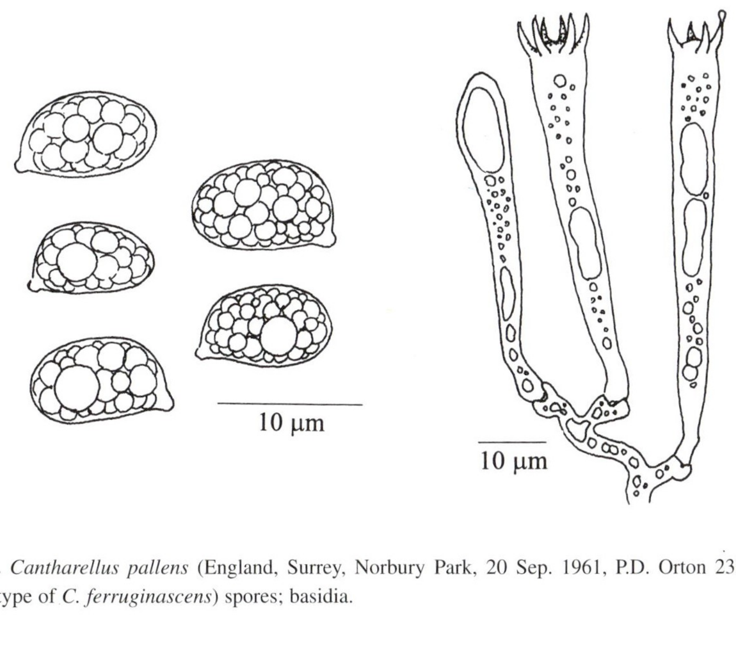

Fig. 11. *Cantharellus pallens* (England, Surrey, Norbury Park, 20 Sep. 1961, P.D. Orton 2331, K(M) 34306, type of *C. ferruginascens*) spores; basidia.

2. AMETHYST CHANTERELLE

Cantharellus amethysteus (Quél.) Sacc., *Syll. Fung*. 5: 482 (1887).
C. cibarius Fr. var. *amethysteus* Quél. in *Assoc. Fr. Avanc. Sci., Congr., Rochelle*: 11 (1882).
Craterellus amthysteus (Quél.) Quél., *Fl. Mycol. Fr.*: 37 (1888).

Pileus 3–6 cm diam., convex to turbinate, finally depressed, surface more or less pale, dull yellow to egg-yellow, initially covered by a greyish violet to pale mauve, felty tomentum which disrupts on expansion to form small, silky squamules or zones over the surface especially at the margin; margin incurved, often plicate. *Hymenophore* arcuate decurrent, 'falsely lamellate', with thick, obtuse ridges, furcate and interveined, anastomosing near the stipe, pale ochraceous yellow to ochre-yellow, often with a slight pinkish tint. *Stipe* 2–4 × 1.5–2.5 cm, obconical, tapering below, firm and solid, surface concolorous with pileus or pale ochraceous, bruising deep yellow to tawny, occasionally pale lilac at apex, flocculose to silky smooth. *Context* up to 1 cm thick, creamy yellow, firm; *generative-hyphae* 2.5–11 µm diam., little inflated, thin-walled, with prominent clamp-connexions; odour and taste faint, pleasant. *Spore deposit* white. *Spores* 8–10(–12) × 4.5–6 µm, ovato-ellipsoid to ellipsoid, rarely slightly constricted, hyaline, thin-walled, with small, refractive granules. Basidia 80–100 × 8–10 µm, elongate clavate, with a basal clamp-connexion; bearing 4–6 sterigmata. *Subhymenial layer* loosely woven, indefinite. *Pileipellis* a loosely woven cutis of somewhat agglutinated hyphae, 4–8 µm diam., with a thickened wall (–1.5 µm) and large clamp-connexions, with or without violaceous vacuolar contents; terminal elements subclavate.

Habitat: in grass, in oak (*Quercus*) and beech (*Fagus*) woods, occasionally with birch (*Betula*) or pine (*Pinus*), often amongst moss.

Distribution and frequency: occasional, throughout northern Europe from August to October.

Other remarks: the fruitbodies occur either singly or in small, subcaespitose clusters. This uncommon species most closely resembles *C. cibarius* mainly differing in the violaceous tints on the pileal surface. Good colour illustrations may be found in Cetto (1987a: 122), Dähncke (1993: 1007), Persson & Mossberg (1994: 44), Ricek (1971). British accounts were provided by Rea (1910: 229, pl. 12; 1922: 542).

Fig. 12. *Cantharellus amethysteus* (England, Kent, Wrotham, Trosley Park, J. Pitt, Sep. 1984; photo. B.M. Spooner).

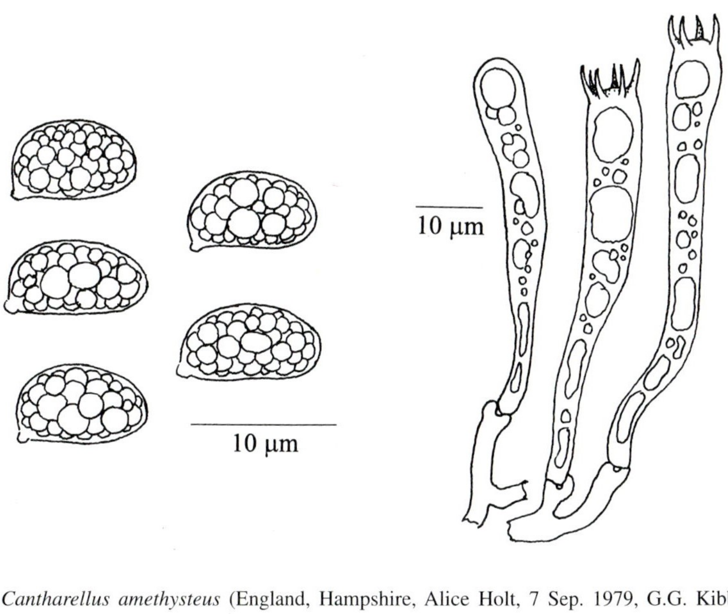

Fig. 13. *Cantharellus amethysteus* (England, Hampshire, Alice Holt, 7 Sep. 1979, G.G. Kibby, K(M) 34340) spores; basidia.

3. THE CHANTERELLE

Cantharellus cibarius Fr., *Syst. Mycol.* 1: 318 (1821).
Agaricus chanterellus L., *Spec. Plant.* 2: 1171 (1753).
Merulius cantharellus Scop., *Fl.. Carn.* ed. 2, 2: 461 (1772).
C. vulgaris Gray, *Nat. Arr. Brit. Pl.* 1: 636 (1821).
Craterellus cibarius (Fr.) Quél., *Fl. Mycol. Fr.*: 37 (1888).

Pileus 3–10(–14) cm diam., fleshy, convex-pulvinate to turbinate when young becoming depressed or funnel-shaped, irregular; surface egg-yellow, orange-yellow to yellow-ochre, at times bruising brownish, glabrous, occasionally rugose; margin lobed, inrolled, and curly. *Hymenophore* deeply decurrent, pseudolamellate, with broad, anastomosing ridges, furcate and interveined, egg-yellow. *Stipe* 3–10 × 0.8–3 cm, massive, generally short, more or less cylindrical, tapering below and merging with the pileus above, concolorous with the pileus, smooth and glabrous. *Context* up to 1.5 cm thick, fleshy-fibrous, whitish to yellowish, drying paler; generative-hyphae 2–6 µm diam., little inflated up to 10 µm diam., hyaline, thin-walled, with prominent clamp-connexions; odour pleasant, of dried apricots; taste mild. *Spore deposit* pale pinkish yellow. *Spores* 7.5–9 × 4–6.5(–7) µm, oblong ellipsoid to ellipsoid, occasionally slightly contricted or apically narrowed, hyaline, thin-walled, with few contents. *Basidia* 60–100 × 7–10 µm, elongate, narrowly clavate, bearing 4–6, rarely 8 sterigmata, and a basal clamp-connexion. *Subhymenial layer* 35–55 µm wide, very indistinct and loosely woven. *Pileipellis* a poorly defined cutis of radially parallel, thinwalled, hyphae, 3–5 µm diam., little differentiated from those of the underlying context.

Habitat: solitary to gregarious, at times in very large numbers, on the ground in both moist, broad-leaved and coniferous woodland, often by the edge of paths. It is considered to have a broad host range, and is found growing especially with beech (*Fagus*), oak (*Quercus*), birch (*Betula*), also spruce (*Picea*) and pine (*Pinus*). It is also known to be associated in Britain with maple (*Acer*), chestnut (*Castanea*), willow (*Salix*) and rowan (*Sorbus*).

Distribution and frequency: common and widespread throughout Europe and North America, but needs much rain and prefers acid soils. June to November.

Other remarks: also known as 'The Girolle'. One of the best known of the edible fungi and sold in many shops and market places throughout Europe. Numerous varieties have been described, and Corner (1966) described fourteen varieties for Europe. The most commonly found in the literature are: var. ***albidus*** R. Maire – with yellowish white hymenophore; var. ***albus*** Fr. – entirely white; var. ***atlanticus*** Romagnesi – pileus with a white pruina, littoral variety, under pines; var. ***bicolor*** R. Maire – basidioma white, hymenophore egg-yellow; var. ***minimus*** L. Daniel – pileus 1.2–1.5 cm diam., and thin-fleshed; subfasciculate in moss; var. ***neglectus*** Souché – pale citrus yellow, hymenophore pale greyish violaceous; var. ***ramosus*** Schulz. – stipe branched; var. ***tenuis*** Romagnesi – small, pileus 2–4 cm diam. subtranslucent, littoral, under pines.

The species is highly prized for its edibility, although its firm texture and delicate flavour require slow, gentle cooking. Throughout Europe the species has been long considered one of the few edible mushrooms, and accounts may be found in the earliest herbals, such as that of Clusius (1601). In Britain, it was illustrated by Bolton (1788: pl. 62) and Sowerby (1797: pl.

46). Good, modern, colour illustrations may be found in Breitenbach & Kränzlin (1986: pl. 481); Comitato di Studio del Gruppo di Vicenza (1989: 214); Marchand (1971: pl. 81); Nilsson & Persson (1978:70–71) and Persson & Mossberg (1994: 36).

The species is well known to form ectomycorrhizal assiociations with the roots of many forest trees, and fruitbodies have not yet been successfully cultivated. It is for this reason that it is harvested to a high degree in the wild, and confusion with similar species can occur. The most common confusion is with the 'False Chanterelle' (*Hygrophoropsis aurantiaca* (Fr.) Maire), a true agaric species of the family *Paxillaceae*. Although harmless, this is tasteless and worthless as an edible species.

Fig. 14. *Cantharellus cibarius* (England, Hampshire, New Forest, Knightwood Incl., Dickson 068/87).

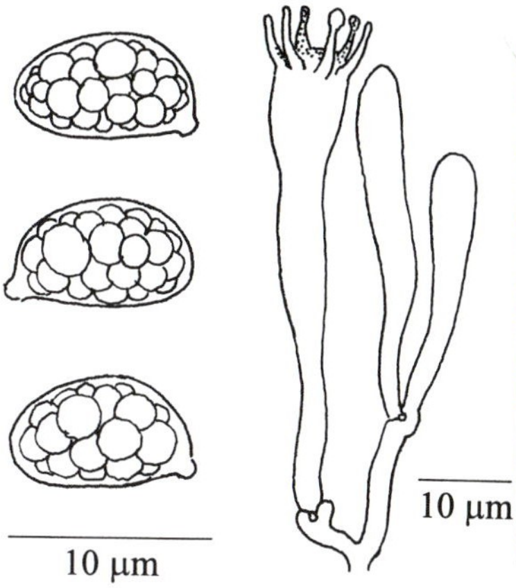

Fig. 15. *C. cibarius* (Scotland, Ross & Cromarty, 8 Jul. 1960, D.A. Reid, K(M) 34652) spores; basidia.

Fig. 16. *Hygrophoropsis aurantiaca* (England, Surrey, Esher Common, N.W. Legon).

4. BLACKENING CHANTERELLE

Cantharellus melanoxeros Desm. apud Duby in DC, *Bot. Gallic.* 2: 799 (1830).

Pileus 2–6(–10) cm diam., fleshy, initially turbinate, expanding to depressed; surface saffron yellow, yellow-ochre to ochre-brown, rarely with a lilaceous tint, blackening towards the margin, more or less radially scrupose; margin thin, lobed and undulate, incurved. *Hymenophore* subdecurrent, with narrow ridges (–1 μm deep), furcate towards the margin, at times anastomosing, rose pink-lilac to grey-violet; edge blunt and rounded. *Stipe* 3–5 × 0.6–1.8 cm, subcylindrical to obconical, often compressed, solid; surface more deeply coloured than pileus, egg-yellow to pinkish lilac, blackening on bruising, glabrous. *Context* pale cream to pale pink, blackening on exposure and with age especially towards the margin and at the stipe base; *generative-hyphae* 2–6(–18) μm diam., hyaline, thin-walled, slightly inflated, with abundant clamp-connexions; *taste* mild, sweet; *odour* fruity. *Spore deposit* cream. *Spores* 7.5–10.0 × 5.0–6.5 μm, ovoid to short ellipsoid, hyaline, thin-walled, smooth, with granular contents. *Basidia* 55–90 × 8–11 μm, elongate clavate, bearing 4–6 sterigmata, with prominent basal clamp-connexion. *Subhymenial layer* 25–35 μm thick, indistinct and loosely woven. *Pileipellis* a thin, disrupted cutis, of radially parallel hyphae, 2.5–5 μm diam., thin-walled, hyaline or occasionally brown.

Habitat: generally forming fused clusters on the soil in beech (*Fagus*) and oak (*Quercus*) woodland.

Distribution and frequency: a species of western and northern Europe. Very rare and unknown from Britain before 1982. Since then collections have been made in Devon, Hampshire, Kent and Inverness-shire.

Other remarks: this species superficially may be mistaken for *C. amethysteus* but is easily distinguished macroscopically by the non-yellow hymenophore. The distinction between *C. melanoxeros* and *C. ianthinoxanthus* (R. Maire) Kühner was established by Monthoux & Röllin (1978), demonstrating the latter species to produce basidiomata with smaller pilei with a thickened margin, a pinkish lilac hymenophore, a pale coloured stipe, and a whitish context which does not blacken on exposure. The spores are also slightly larger, measuring '9.4–11 × 6.6–7.3 μm'. The fungus described and illustrated by Damblon, Demoulin & Schumacher (1974 & 1975: pl. 199) and Breitenbach & Kränzlin (1986: pl. 484) represents *C. ianthinoxanthus*. The account by Persson & Mossberg (1994: 60) emphasizes the blackening context and the bright yellowish stipe of *C. melanoxeros*. Good colour illustrations may also be found in Cetto (1987a: 126), Comitato di Studio del Gruppo di Vicenza (1989: 218) and Monthoux & Röllin (1978). *Craterellus incarnatus* Quél. (= *Cantharellus ciliatus* Corner, non *Cantharellus incarnatus* (Beeli) Heinem.) is often listed as a synonym but, according to Monthoux & Röllin (1978), represents neither *C. ianthinoxanthus* nor *C. melanoxeros* but another species which forms a white basidioma when young. Fiasson (1973) has demonstrated the presence of aliphatic carotenes, indicating that this group of species belongs to the subgenus *Phaeocantharellus*.

Fig. 17. *Cantharellus melanoxeros* (England, Devon, Loddiswell, Andrew's Wood, Oct. 1995, S. Evans).

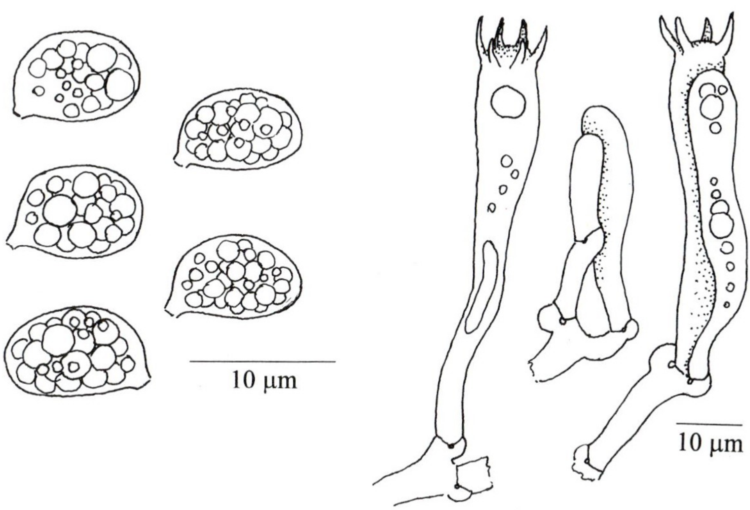

Fig. 18. *Cantharellus melanoxeros* (England, Devon, Loddiswell, Andrew's Wood, 16 Oct. 1995, D. Shorten, K(M) 31579) spores; basidia.

5. YELLOW-STEMMED CHANTERELLE

Cantharellus aurora (Batsch) Kuyp. in *Riv. Micol.* 33: 249 (1991).
Agaricus aurora Batsch, *Elench. Fung.*: 93 (1783).
Merulius xanthopus Pers., *Mycol. Eur.* 2: 19 (1825).
M. auroreus Pers.. *loc. cit.*: 19
M. cervinus Pers., *loc. cit.*: 20
Cantharellus xanthopus (Pers.) Duby in DC, *Bot. Gallic.* 2: 799 (1830).
C. lutescens sensu Fr., *Syst. Mycol.* 1: 320 (1821), non *Merulius lutescens* Pers., *C. tubiformis* var. *lutescens* Fr.

Pileus 2–6 cm diam., thin-fleshed, convex-umbilicate then depressed to funnel-shaped on expansion, often perforate; surface grey-brown to almost blackish when young, then orange-brown, paler at the margin, radially fibrillose, wrinkled, floccose-scaly towards the margin; margin crispate, strongly undulate. *Hymenophore* short decurrent, smooth, finally slightly veined-wrinkled, with shallow, well spaced ridges towards the margin, bright yellowish orange near the stipe becoming greyish brown towards the pileal margin. *Stipe* 5–8 × 0.5–1.5 cm, sinuous cylindrical, often compressed, tapering below, pitted, hollow; surface golden yellow to saffron-yellow, at times with salmon-pink tints, paler at base, smooth, usually with a white, tomentose base. *Context* very thin, soft-fibrous, cream; *generative-hyphae* 3–8(–18) µm diam., hyaline, thin-walled, frequently inflating, with prominent clamp-connexions; odour strong, fruity (apples, apricots); taste mild. *Spore deposit* cream. *Spores* 8.5–12.5 × 6.5–8.5 µm, ovoid to broadly ellipsoid, hyaline, thin-walled, with vacuolar contents. *Basidia* 60–90 × 9–12 µm, elongate clavate, bearing (2–)4 sterigmata, with a basal clamp-connexion. *Subhymenial layer* 20–30 µm thick, loosely woven. *Pileipellis* a repent cutis, of subagglutinated, thin-walled hyphae, 3–9 µm diam., with yellow-brown vacuolar contents.

Habitat: gregarious, forming large clusters, in mossy pine- and spruce-woods (*Pinus* and *Picea*), often at edge of swamps, amongst *Sphagnum* moss, preferring alkaline, calcareous soils.

Distribution and frequency: uncommon throughout Europe and eastern North America. Rare in Britain, although more common in Scotland and Ireland (Cork and Wicklow), where it is confined to mountainous areas. August to September.

Other remarks: a rarely seen species, and limited to high altitude regions throughout northern Europe. It differs macroscopically from *C. tubiformis* by the almost smooth hymenophore, which only forms widely spaced, very low ridges, never developing a pseudolamellate condition. This is the species widely known under the name *Cantharellus lutescens* (Pers.) Fr. When Fries (1821: 320) described the above species, with a smooth to vein-like hymenophore, he misapplied the name *Merulius lutescens* Pers., which refers to a taxon with a pseudolamellate hymenophore, typical of *C. tubiformis*. This confusion led Donk (1969: 273) to select *Merulius xanthopus* Pers. as the basionym and accept it as the correct name for *C. lutescens* sensu Fr. However, Kuyper (1990) pointed out that *Agaricus aurora* Batsch was the earliest name for the fungus, and introduced the combination of *Cantharellus aurora*. This name was utilized by Persson & Mossberg (1994: 54), who provided an excellent illustration. Additional colour illustrations may be found in Breitenbach & Kränzlin (1986: pl. 483), Cetto (1987: 130), Dähncke (1993: 1011), Comitato di Studio del Gruppo di Vicenza (1989: 210).

Fig. 19. *Cantharellus aurora* (Scotland, Perthshire, Black Wood of Rannoch, Sept. 1982, G. Dickson 731/82).

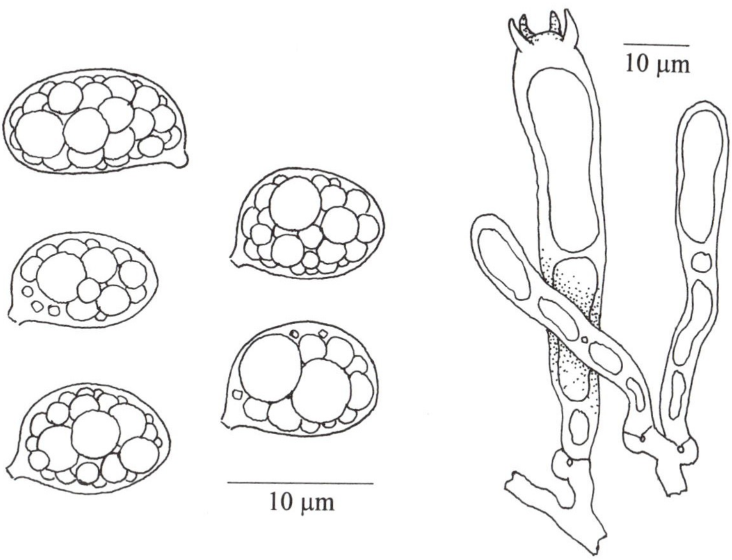

Fig. 20. *Cantharellus aurora* (Scotland, Inverness-shire, Loch-an-Eilean, 31 Aug. 1955, D.A. Reid, K(M) 34783) spores; basidia.

6. ORANGE CHANTERELLE

Cantharellus friesii Quél., *Champ. Jura Vosges* 1: 191 (1872).
Craterellus friesii (Quél.) Quél., *Fl. Mycol. Fr.*: 38 (1888).
Cantharellus miniatus Fayod in *Ann. R. Accad. Agric. Torina* 35: 82 (1893).

Pileus 1–4.5 cm diam., convex soon expanding to depressed and almost funnel-shaped with an umbilicate centre; surface deep orange to pinkish orange, yellowish orange at the centre, paler when mature and at margin, smooth to finely furfuraceous, glabrescent with silky sheen; margin undulate, inrolled. *Hymenophore* subdecurrent, arcuate, pseudolamellate, with irregularly furcate and anastomosing, narrow ridges, pale flesh-pink or pinkish lilaceous, then pale orange to pale yellow-orange. *Stipe* 1–4 × 0.2–0.9 cm, more or less cylindrical, merging with the pileus, at first solid then hollow when old; surface concolorous with pileus or yellowish orange, often with darker apex, silky smooth, whitish towards the base. *Context* thin, fibrous-fleshy, concolorous, drying white to yellowish; generative-hyphae 2.5–6 µm diam., inflating to 17 µm diam., hyaline, thin-walled, with prominent clamp-connexions; odour none to slight; taste sharp, acidic. *Spore deposit* pale cream. *Spores* 8.5–12 × 4.5–7.5 µm, ovoid to ellipso-cylindrical, hyaline, thin-walled, smooth, guttulate. *Basidia* 55–90 × 7–10 mm, elongate clavate, bearing 2–5 sterigmata. *Subhymenial layer* 20–30 µm thick, loosely woven. *Pileipellis* hyphae a disrupted, repent cutis of radially parallel, subagglutinated hyphae, 3–10 µm diam., containing an orange vacuolar pigment.

Habitat: solitary or in small groups, on bare or moss-covered soil, often under beech (*Fagus*), also birch (*Betula*).

Distribution and frequency: prefers northerly or mountainous regions in Europe. Rare in the British Isles, probably confined to northern localities.

Other remarks: this species resembles a small version of The Chanterelle, but is easily separated by the pale pinkish hymenophore, and longer spores. Descriptions of British material are provided by Orton (1964) and Rea (1922: 542). Colour illustrations may be found in Breitenbach & Kränzlin (1986: pl. 482); Bresadola (1929: pl. 472); Cetto (1987a: 124), Comitato di Studio del Gruppo di Vicenza (1989: 217); Dähncke (1993: 1010), Persson & Mossberg (1994: 48) and Quélet (1872: pl. 23/2).

Fig. 21. *Cantharellus friesii* (Scotland, Perthshire, Glen Lyon, P.D. Orton).

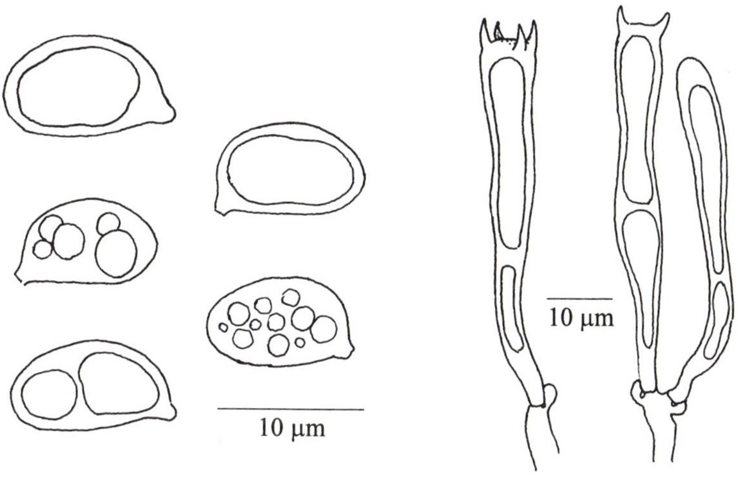

Fig. 22. *Cantharellus friesii* (Scotland, Perthshire, Glen Lyon, 30 Sep. 1961, P.D. Orton, E) spores; basidia.

7. AUTUMN CHANTERELLE

Cantharellus tubiformis Bull.: Fr., *Syst. Mycol.* 1: 319 (1821, as *tubaeformis*).
Peziza undulata Bolton, *Hist. fung. Halifax*: pl. 105/2 (1789).
Helvella tubaeformis Bull., *Herb. Fr.*: pl. 461 (1789).
H. cantharelloides Bull., l. c.: pl. 473/3. (1789).
Agaricus cantharelloides Bull., l. c.: pl. 505/2 (1790), non Sowerby (1797).
Merulius tubiformis (Bull.) Pers., *Synops. Meth. Fung.* 1: 489 (1801).
M. lutescens Pers., l. c.: 489
M. villosus Pers., *Mycol. Eur.* 2: 14 (1825).
Craterellus lutescens (Pers.) Fr., *Epicrisis*: 532 (1838).
Craterellus tubiformis (Bull.: Fr.) Quél., *Fl. Mycol.*: 36 (1888).
Craterellus cantharelloides (Bull.) Quél., *Ass. Fr.*: 5 (1895).
Cantharellus infundibuliformis (Scop.) Fr. var. *tubaeformis* R. Maire in *Mus. Barcin. Sci. Nat. Op.* 15: 44 (1933).

7a. var. **tubiformis**

Pileus 1.5–6 cm diam., thin-fleshy, convex then depressed, umbilicate then pervious and more or less funnel-shaped; surface greyish yellow, olivaceous yellow, ochre-grey to grey-brown, smooth when young soon becoming wrinkled and scaly around the disk, with a curly, irregular margin. *Hymenophore* decurrent, pseudolamellate, with distinct, thick ridges, interveined in large specimens, pale yellowish, at times with greenish tint, becoming pale grey and tinged colour of pileus. *Stipe* 1.5–8 × 0.3–1 cm, cylindrical but becoming compressed, tapering below, at times grooved or pitted, hollow; surface greyish yellow, fuliginous yellow or grey, dry, glabrous, silky, fibrilloso-striate. *Context* thin, whitish to sepia yellow; *generative-hyphae* 2.5–7(–15) µm diam., hyaline, thin-walled, inflated, with clamp-connexions; *odour* pleasant; *taste* mild. *Spore deposit* creamish white to pale lemon yellow. *Spores* 9–12.5 × 7–10 µm, subglobose to broadly ovoid, hyaline, thin-walled, with few contents. *Basidia* 60–100 × 9–12 µm, elongate clavate, bearing (2–)4–5 sterigmata, with basal clamp-connexion. *Subhymenial layer* 20–45 µm indistinct and loosely woven. *Pileipellis* a repent cutis of radially parallel, slightly agglutinated hypahe, 4–11 µm diam., thin-walled, with brown, vacuolar contents.

Habitat: in troops in both coniferous and deciduous woodland.

Distribution and frequency: an autumn fungus, found throughout Europe and northern North America. September to November, can survive early frosts. A fairly common species but much more so in northern localities where it may be found in enormous numbers.

Other remarks: the species can grow in very large numbers covering a wide area of woodland, and can also survive early frosts, unlike related species. It is therefore collected on a commercial basis throughout Scandinavia. Edibility is regarded by some as superior to that of The Chanterelle, and the fruitbodies are often preserved by drying. The fungus, however, may be difficult to find at first, as the fruitbodies blend with the fallen, withered leaves. Sometimes fruitbodies are found with entirely grey pigments, lacking yellow or orange, and these may be confused with *Pseudocraterellus cinereus*. Good colour illustrations are to be found in Breitenbach & Kränzlin (1986: pl. 485); Cetto (1987a: 134), Comitato di Studio del

Gruppo di Vicenza (1989: 220); Dähncke (1993: 1012); Marchand (1973: pl. 177); Nilsson & Persson (1978: 72–73); Persson & Mossberg (1994: 52).

7b. var. **lutescens** Fr., *Epicrisis*: 366 (1838).
Merulius infundibuliformis Scop., *Fl. Carn.* ed. 2, 2: 462 (1772).
Agaricus cantharelloides Sow., *Col. Fig. Engl. Fung.* 1: pl. 47 (1797), non Bulliard (1790).
Cantharellus infundibuliformis (Scop.) Fr., *Epicrisis*: 366 (1838).

Pileus surface pale yellow or more rarely greyish brown. *Hymenophore* and stipe pale yellow. All other characters similar to those of var. *tubiformis*.

Habitat, distribution and frequency: as var. *tubiformis*.

Other remarks: this variety is the more common in Britain, resembling the typical form but having a yellow stipe and yellow hymenophore ridges. British authors, such as Cooke (1890: pl. 1061 (1108)) and Dennis, Orton & Hora (1960) have applied the name *C. infundibuliformis* for *C. tubiformis* var. *lutescens*. The distinction between *C. tubiformis* and *C. infundibuliformis* has been long disputed. Fries (1838) separated *C. infundibuliformis* largely on the basis of a yellow rather than an orange tawny stipe, and this was maintained by Corner (1966: 71). On the other hand, Donk (1933: 7) maintained *C. infundibuliformis* as a synonym of *C. tubiformis*. Fries (1838: 366) emphasized the clear yellow ('flavus') stipe when he described *C. infundibuliformis*, whilst at the same time citing the plate published by Sowerby (1797: pl. 47) to be the best illustration. Sowerby, although giving the name '*Agaricus cantherelloides*', distinguished the English fungus, on the basis of the yellow stipe, from the fungus of the same name published earlier by Bulliard (1789: pl. 505/2) which has a dark stipe. Dähncke (1993: 1013) is a good illustration of this variety.

Fig. 23. *Cantharellus tubiformis* v. *lutescens* (England, Shropshire, Wyre Forest, Brand Wood, 11 Sep. 1924, C. Rea) from an original painting by the collector.

Fig. 24. *Cantharellus tubiformis* v. *tubiformis* (England, Hampshire, New Forest, Denny Wood, N.W. Legon).

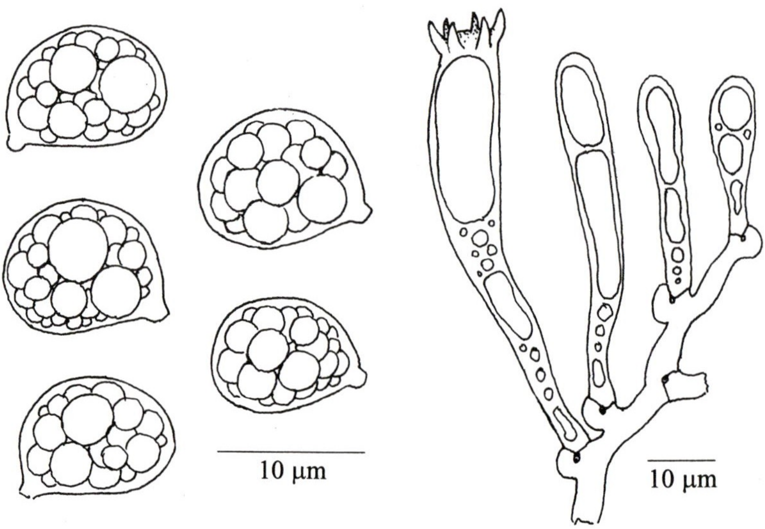

Fig. 25. *Cantharellus tubiformis* v. *tubiformis* (Scotland, Perthshire, Black Wood of Rannoch, 24 Sep. 1983, D.N. Pegler 3569, K(M) 34845) spores; basidia.

3. HYDNACEAE F. F. Chev.

Fl. gén. env. Paris : 270 (1826).

Fruitbody pileate and stipitate. *Pileus* fleshy, white, yellow or orange. *Hymenophore* spinose; spines terete, pointed, concolorous with or paler than pileus. *Stipe* well developed, central or excentric. *Veils* absent. *Context* fleshy, brittle, white to cream; *hyphal system* monomitic; generative-hyphae thin-walled, inflating, with clamp-connexions. *Spore deposit* white to pale yellow. *Basidiospores* subglobose to ellipsoid, hyaline, inamyloid, non-dextrinoid, smooth, thin-walled. *Basidia* clavate, stichic, bearing 2–4 sterigmata. *Cystidia* not developed. Gymnocarpic. Terricolous or humicolous. World-wide. *Type genus*: *Hydnum* L.: Fr.

HYDNUM L.: Fr.

Syst. Mycol. 1: 397 (1821); Linn., *Spec. Plant.* 2: 1178 (1753).

Hypothele Paul., *Mycétol.*: 43 (1812).
Dentinum Gray, *Nat. arrang. Brit. Pl.* 1: 650 (1821).
Tyrodon P. Karst. in *Rev. Mycol.* 3 (9): 19 (1881).

Pileus pulvinate then depressed, whitish to orange-brown, finely tomentose, glabrescent, rarely scaly. *Hymenophore* free to decurrent, spinose; *spines* white or concolorous with pileus. *Stipe* well developed but short and stocky, solid; surface concolorous or paler, finely tomentose, glabrescent. *Context* fleshy, not zoned; *hyphal-system* monomitic; *generative-hyphae* inflating, clamp-connexions present. *Spore deposit* white to ochraceous. *Spores* subglobose to obovoid, hyaline, smooth, thin-walled, inamyloid. *Basidia* narrowly clavate, 2–4, rarely more sterigmate. *Hymenophoral trama* regular. Terricolous. *Type species*: *H. repandum* L.: Fr.

Key to Species

1. Fruitbody large and robust (Fig. 26); pileus with pale colours; stipe usually excentric; hymenophore often subdecurrent .... - **1. Wood Urchin** (*H. repandum*)
1. Fruitbody smaller and slender (Fig. 28); pileus deeply coloured; stipe typically central; hymenophore free to adnate, not decurrent ..... - **2. Reddish Wood Urchin** (*H. rufescens*)

1. WOOD URCHIN

Hydnum repandum L.: Fr., *Syst. Mycol.* 1: 400 (1821); L., *Spec. Plant.* 2: 1178 (1753).
Hydnum carnosum Batsch, *Elench. Fung.*: 111 (1783).
H. rufescens Schaeff., *Fung. Icon.* 4: 95 (1774), non Fr. (1821).
Dentinum repandum (L.: Fr.) Gray, *Nat. arrang. Brit. Pl.* 1: 650 (1821).
Tyrodon repandus (L.: Fr.) P. Karst. in *Rev. Mycol.* 3 (9): 19 (1881).
Sarcodon repandus (L.: Fr.) Quél., *Elench. Fung.*: 189 (1886).
Hypothele repanda (L.: Fr.) Banker in *Torreya* 4: 113 (1904).
Hydnum heimii Maas G. in *Persoonia* 1: 133 (1959).

Fruitbody solitary or concrescent. *Pileus* 5–10(–17) cm diam., at first convex, variable, pulvinate then applanate to depressed; surface white, cream to pale pinkish buff, tomentose, velvety to felty, glabrescent, smooth and shiny when dry, not zoned; margin entire, thick, persistently inrolled, splitting, finally lobed or undulate. *Hymenophore* spinose, adnate, at times subdecurrent at least down one side of the stipe, white or with a pinkish tint; *spines* subulate, soft-brittle, fragile, uneven in length, 4–8 mm long × 2–3 mm diam., very crowded. *Stipe* 2–7 × 1.5–4 cm, central but mostly excentric, fairly stocky, cylindrical or swollen below; solid; surface whitish, paler than pileus, finely tomentose, glabrescent, smooth, arising from an abundant mycelial mass. *Context* thick, soft-fleshy, white, occasionally discolouring reddish on bruising; *generative-hyphae* 3–5 µm, inflating to 25 µm diam., hyaline, thin-walled, with clamp-connexions; *taste* mild to bitter; *odour* pleasant but faint. *Spore deposit* pale cream. *Spores* 6.5–8.5 × 5.5–7 µm, subglobose to broadly ovoid, hyaline, thin-walled, smooth,usually containing a single, large, refractive guttule. *Basidia* 30–45 × 6–10 µm, short clavate, bearing 4 sterigmata, and a basal clamp-connexion. *Subhymenial layer* 10–20 µm wide, interwoven. *Hymenophoral trama* regular, with narrow, uninflated, thin-walled hyphae, 2–5 µm diam., with prominent clamp-connexions. *Pileipellis* a disrupted and indefinite trichodermium, of narrowly clavate elements, 2.5–4.5 µm diam., with a pale yellowish wall.

Habitat: amongst humus in deciduous woods, especially oak (*Quercus*), beech (*Fagus*), and birch (*Betula*), where it can be abundant, but also found in pine and spruce woodland (*Pinus* and *Picea*).

Distribution and frequency: common throughout Europe and North America during August to November. Solitary to gregarious.

Other remarks: this paler species is larger and more fleshy than *H. rufescens*. Generally with a pleasant odour, the edibility is comparable to that of the Chanterelles, although it can be bitter if eaten raw. Also known as the 'Hedgehog Fungus', the 'Wood Hedgehog', and commercially as the 'Pied de Mouton'. The full and complex synonymy of this species may be found in Mass Geesteranus (1959: 134). Good illustrations in Maas Geesteranus (1975: pl. 7–8); Marchand (1973: pl. 181); Ryman & Holmåsen (1984: 107). Occasionally, a pure white form, *forma* ***albidum*** (Peck) Nikol., is encountered, which tends to be more robust and of firmer texture, resembling *Russula delica* Fr. from above.

Fig. 26. *Hydnum repandum* (England, Hampshire, New Forest, Denny Wood, N.W. Legon).

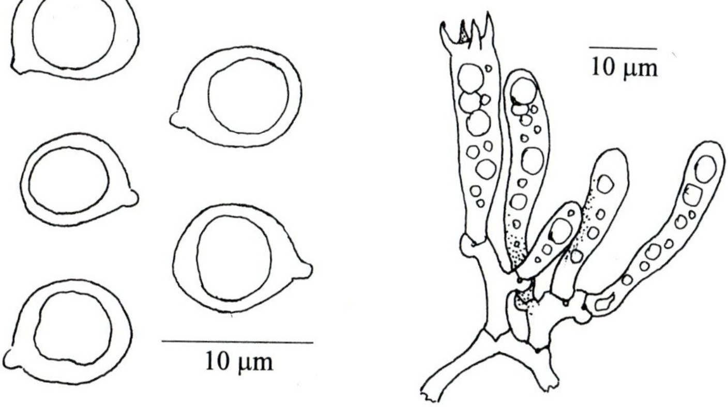

Fig. 27. *Hydnum repandum* (England, Surrey, Esher Common, 27 Sep. 1994, N.W. Legon, K(M) 27566) spores; basidia.

2. REDDISH WOOD URCHIN

Hydnum rufescens Fr., *Syst. Mycol.* 1: 401 (1821), non *H. rufescens* Schaeff. (1774).
H. repandum var. *rufescens* (Fr.) Barla, *Champ. Prov. Nice*: 81 (1859).
Sarcodon repandus (L.: Fr.) Quél. var. *rufescens* (Fr.) Quél., *Elench. Fung.*: 189 (1886).
Tyrodon rufescens (Fr.) P. Karst. in *Bidr. känn. Finl. Nat. Folk* 48: 349 (1889).
Dentinum rufescens (Fr.) Pouz. in *Ceská Mykol.* 10: 76 (1956).

Pileus 3–7 cm diam., convex to applanate, then depressed in centre; surface ochraceous orange to bright fulvous orange, drying paler from the margin, velvety, glabrescent, rarely scaly; margin involute then straight. *Hymenophore* free to adnate, not truly decurrent, spinose; *spines* up to 5 mm long × 2–3 mm, yellow-orange to salmon pink, moderately crowded. *Stipe* usually central, 2–7 × 0.5–1.5 cm, slender to stocky, solid; white or with pinkish tints, discolouring deep yellow when bruised, tomentose, glabrescent. *Context* soft-fleshy, pale flesh pink, discolouring yellow on exposure; generative-hyphae 2–5 µm diam., inflating up to 20 µm diam., thin-walled, with clamp-connexions; *taste* mild. *Spores* 7.5–10 × 6.5–8.5 µm, subglobose to broadly ovoid, hyaline, smooth, thin-walled, containing large, refractive oil drops. *Basidia* 35–45 × 7–10 µm, clavate, bearing 2 or 4 sterigmata, with basal clamp-connexion. *Subhymenial layer* 10–15 µm wide, interwoven. *Hymenophoral trama* regular, with narrow, uninflated, thin-walled hyphae, 2–5 µm diam., with prominent clamp-connexions. *Pileipellis* a disrupted and indefinite trichodermium, of narrowly clavate elements, 2.5–4.5 µm diam., with a pale yellowish to brown wall.

Habitat: usually in damper situations than *H. repandum*. More often associated with beech (*Fagus*), but also maple (*Acer*), birch (*Betula*), hazel (*Corylus*), and oak (*Quercus*).

Distribution and frequency: uncommon but widespread throughout Europe and North America. August to November.

Other remarks: *Hydnum rufescens* is often regarded as a variety of *H. repandum*, and was so treated by Rea (1922: 631) and Wakefield & Dennis (1950: 247), but the two taxa are separated by most north European authors, such as Maas Geesteranus (1975: 87). Microscopically, the high percentage of bisporic basidia would appear to be a distinctive character for the species. Good illustrations may be found in Maas Geesteranus (1975: pl. 9); Ryman & Holmåsen (1984: 107).

Fig. 28. *Hydnum rufescens* (England, Hampshire, New Forest, Whitley Wood, Oct. 1970, G. Dickson 279/70).

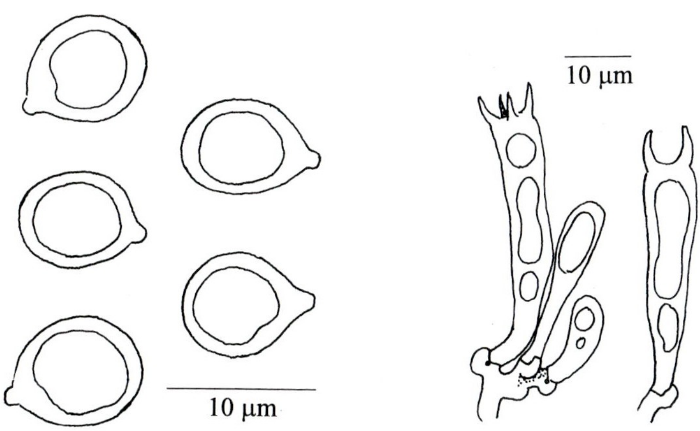

Fig. 29. *Hydnum rufescens* (England, Surrey, Witley Common, 18 Sep. 1977, E. Chambers, K(M) 35166) spores; basidia.

SPINE FUNGI

HERICIALES Jülich
in *Biblio. Myc.* 85: 348 (1982)

Fruitbody sessile or stipitate, effused, pileate, or clavarioid. *Hymenophore* smooth, toothed, poroid, or lamellate. *Context* fleshy to tough. *Hyphal system* monomitic or dimitic, *generative-hyphae* amyloid or inamyloid, with or without clamp-connexions. *Spores* ballistosporic, hyaline, normally strongly amyloid (grey to violaceous in Melzer's reagent), smooth or punctate. *Basidia* clavate to suburniform, bearing (2–)4 sterigmata. *Gloeoplerous elements* normally present, often emerging in the hymenium as long gloeocystidia. Lignicolous, more rarely terricolous, muscicolous, or litter-inhabiting. Worldwide. *Type family: Hericiaceae* Donk.

The *Hericiales* are a diverse, possibly polyphyletic group of mostly wood-rotting fungi typified by their gloeoplerous (oil-filled) hyphae and amyloid spores. Five families are recognized: *Auriscalpiaceae*, *Clavicoronaceae*, *Gloeocystidiellaceae*, *Hericiaceae*, and *Lentinellaceae*; but only those with hydnoid representatives in the British Isles (*Auriscalpiaceae* and *Hericiaceae*) are dealt with in this book. The *Lentinellaceae* are pileate and agaricoid (mushroom-like), with several British species growing on wood, superficially resembling Oyster Mushrooms. The *Clavicoronaceae* produce small, clavarioid (club-shaped) or coral-like fruitbodies; only two rare species occur in Britain. The *Gloeocystidiellaceae* are effused fungi forming patches on the underside of dead wood with a number of fairly common British representatives. Though the fruitbody shapes are extremely diverse, all these families share a similar micromorphology, suggesting they have evolved from a common ancestor.

Key to British Hydnoid Families

1. Fruitbodies rather tough, often dimitic, effused to pileate. Single British species pileate with long thin stipe, arising from fallen conifer cones .... 4. **Ear Pick-Fungi** (*Auriscalpiaceae*)
1. Fruitbodies (in hydnoid genera) soft, always monomitic, consisting of isolated pendent spines, or large, sometimes compound, bracket-like growths with long pendent spines or teeth, on dead wood .... 5. **Spine & Prickle Fungi** (*Hericiaceae*)

4. **AURISCALPIACEAE** Maas Geest.

in *Proc. K. Ned. Akad. Wet.* (Ser. C) 66: 426 (1963).

Fruitbody rather tough, whitish to dark brown, effused or pileate, sessile or stipitate. *Hymenophore* hydnoid or poroid. *Spores* globose to ellipsoid, hyaline, amyloid, smooth to finely echinulate. *Basidia* clavate to suburniform, bearing (2–)4 sterigmata. *Hyphal system* monomitic or dimitic; *generative-hyphae* hyaline to brown, with or without clamp-connexions. *Gloeoplerous elements* present. Lignicolous, terricolous, or muscicolous. Worldwide. *Type genus*: *Auriscalpium* Gray.

The family is close to the *Hericiaceae* but has been considered distinct in often (though not always) having tough, leathery fruitbodies which in many species are dimitic, containing thick-walled skeletal hyphae. Five genera are recognized. Four of these, *Amylonotus* Ryvarden, *Amylosporus* Ryvarden, *Dentipellis* Donk, and *Steccherinum* D.A. Reid, are effused or bracket-like fungi not represented in the British Isles. The fifth genus, *Auriscalpium*, has a single British species.

AURISCALPIUM Gray

Natural Arrang. Brit. Plants 1: 650 (1821).

Pleurodon P. Karst. in *Rev. mycol.* 3(9): 20 (1881).

Fruitbody solitary or partly confluent, pileate, arising from a lateral stipe, smooth to pilose. *Hymenophore* hydnoid. *Context* soft to leathery, inamyloid. *Hyphal system* monomitic or dimitic; *generative hyphae* hyaline to brown, not inflated, with clamp-connexions. *Spores* ovoid to ellipsoid, hyaline, amyloid, smooth to finely punctate or verrucose. Basidia clavate, bearing 4 sterigmata. *Gloeoplerous elements* conspicuous to inconspicuous, sometimes appearing as gloeocystidia in the hymenium. Lignicolous, terricolous, or muscicolous. Worldwide. *Type species*: *Auriscalpium vulgare* Gray.

The genus is distinguished by its stipitate, hydnoid fruitbodies and amyloid spores. Maas Geesteranus (1978) provided a key to world species. There is a single species in Europe.

EAR PICK-FUNGUS

Auriscalpium vulgare Gray, *Natural Arrang. Brit. Plants* 1: 650 (1821).
Hydnum auriscalpium L., *Spec. Pl.* 2: 1178 (1753).
Pleurodon auriscalpium (L.) P. Karst. in *Rev. mycol.* 3(9): 20 (1881).

Fruitbody pileate, up to 20 mm diam., thinly or densely covered with hairs, becoming glabrous with age; pale or pinkish brown, darkening with age. *Hymenophore* hydnoid, with teeth or spines up to 3 mm long, paler than the pileus. *Stipe* lateral, up to 60 × 2 mm, straight or flexuous, thinly or densely covered with hairs, becoming glabrous with age, mid to dark brown. *Context* rather tough, leathery, whitish. *Hyphal system* dimitic; *generative-hyphae* 2–3 µm diam., not inflated, thin-walled and hyaline in context and hymenium, thick-walled and brown at pileal and stipe surface, inamyloid, with clamp-connexions; *skeletal hyphae* 2–3 µm diam., thick-walled to solid, hyaline in context, brown in mass in trama of spines. *Hairs* on pileus and stipe composed of agglutinated, brown, thick-walled, generative hyphae, often with unclamped secondary septa. *Spores* 4.5–6.5 × 3.5–5 µm, ellipsoid, hyaline, amyloid, finely punctate, hilar appendix inconspicuous. *Basidia* 16–22 × 4–6 µm, clavate, bearing 4 sterigmata. *Gloeoplerous elements* absent in context, but present as inconspicuous and often sparse gloeocystidia in the hymenium.

Habitat: arising from rotten or buried *Pinus* cones. Occasionally reported on *Picea* cones.

Distribution and frequency: widespread and common throughout the British Isles and continental Europe. Throughout the year. Also found in North America.

Other remarks: the only European representative of its genus, *Auriscalpium vulgare* is distinctive and easy to recognize in the field. At first sight, fruitbodies often appear to be growing out of the ground or needle layer, but careful search will show that they are attached to buried cones or cone debris.

Additional colour photographs may be found in Breitenbach & Kränzlin (1986: pl. 283), Dähncke (1993: 1038), Marchand (1976: pl. 334), Moser & Jülich (1985: V *Auriscalpium* 1, 2, & 3), and Ryman & Holmåsen (1984: 108).

Fig. 30. *Auriscalpium vulgare* (England, Oxfordshire, Chiltern Hills, 12 Sep. 1969, A.W. Brand 1556).

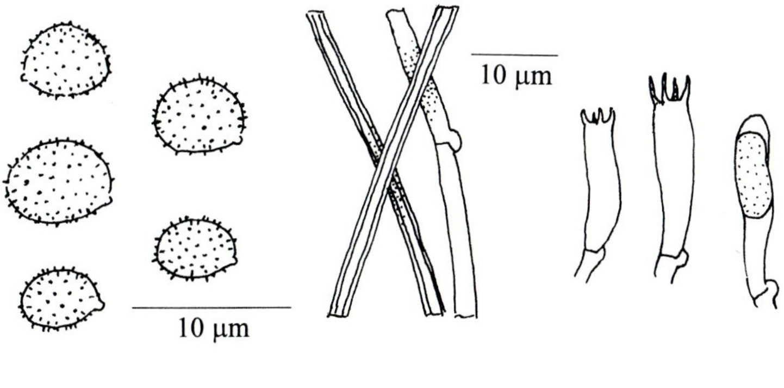

Fig. 31. *Auriscalpium vulgare* (England, Surrey, Esher Common, 6 Oct. 1994, Y.J. Yao, K(M) 27497) spores; thick-walled skeletal hyphae and thin-walled, clamped, generative hyphae; basidia and gloeocystidium.

5. HERICIACEAE Donk
in *Persoonia* 3: 269 (1964).

Fruitbody (in hydnoid genera) soft, whitish or pale, largely or entirely consisting of pendent spines or teeth, sessile or bracket-like, sometimes stalked or branched. *Hymenophore* of pendent spines or teeth. *Context* soft. *Hyphal system* monomitic; *generative-hyphae* hyaline, inflated or non-inflated, thin to thick-walled, amyloid or inamyloid, normally with clamp-connexions. *Spores* globose to ellipsoid, smooth to punctate, hyaline, amyloid. *Basidia* clavate, bearing (2–)4 sterigmata. *Cystidia* absent or present. *Gloeoplerous elements* absent or present, sometimes emerging in the hymenium as gloeocystidia. Lignicolous or litter-inhabiting. Worldwide. *Type genus*: *Hericium* Pers.

Two hydnoid genera are known from the British Isles.

Key to British Hydnoid Genera

1. Fruitbody small, inconspicous, consisting of discrete pendent spines arising directly from the substratum .... 1. **Mucronella**
1. Fruitbody large, conspicuous, bracket-like, with pendent spines connected to sterile tissue .... 2. **Hericium**

1. MUCRONELLA Fr.
Hymen. eur. 629 (1874)

Fruitbody & hymenophore composed of numerous individual teeth or spines, 0.5–8 mm long, directly pendent from the underside of fallen branches and logs, without or with only a very thin subiculum. *Individual spines* discrete or fasciculate, normally attached by a short, inconspicuous, sterile stalk. *Context* soft, fleshy. *Hyphal system* monomitic; *generative-hyphae* not inflated, hyaline, inamyloid, thin-walled, with clamp-connexions. Spores globose to ellipsoid, hyaline, amyloid, smooth to punctate. *Basidia* clavate, bearing 4 sterigmata. *Cystidia* present or absent. *Gloeoplerous elements* present or absent. Lignicolous or litter-inhabiting. Worldwide. *Type species*: *Mucronella calva* (Alb. & Schwein.) Fr.

Most *Mucronella* species lack gloeoplerous elements, but otherwise seem related to the *Hericiaceae* in their overall microstructure and in their possession of amyloid spores. Corner (1970) provided a worldwide key. A single species in the British Isles.

PRICKLE FUNGUS

Mucronella calva (Alb. & Schwein.) Fr., *Hymen. Eur.*: 629 (1874).
Hydnum calvum Alb. & Schwein., *Consp. Fung.*: 271 (1805).
Mucronella aggregata Fr., *Monogr. Hymen. Suec.* 2: 280 (1863).

Fruitbody & hymenophore composed of numerous individual teeth or spines, 0.5–5 mm long, directly pendent from the underside of rotten fallen branches and logs, without subiculum or visible connective tissue; each individual spine is attached by a short sterile stem, often reduced and inconspicuous, and is acutely tapered from stem to apex; collective fruitbodies may comprise several hundred spines; spines white to yellow, soft, fleshy, easily detached. *Context* soft, fleshy. *Hyphal system* monomitic; *generative-hyphae* 2.5–5(–8) µm wide, not inflated, hyaline, inamyloid, thin-walled, with clamp-connexions. *Spores* 4–5(–6) × 2.5–3 µm, ellipsoid, amyloid, hyaline, smooth to finely punctate. *Basidia* 15–20 × 4–5 µm, clavate, bearing 4 sterigmata. *Gloeoplerous elements* absent.

Habitat: on underside of fallen deciduous or coniferous branches, trunks, and logs. Less commonly in leaf litter or tiered on rotten stumps.

Distribution and frequency: widespread and common in southern England, but only one specimen in K from northern England (Yorkshire) and one from Scotland (Galloway). Throughout the year. Widespread and common throughout continental Europe.

Other remarks: *Mucronella calva* is inconspicuous compared to the other fungi treated in this book, but readily recognizable in the field by its unusual growth habit. Many common corticioid (patch-forming) fungi have a toothed hymenium and are also found under logs, but in all cases the teeth are attached to a visible subiculum.

Mucronella aggregata is a name sometimes given to densely growing forms of *M. calva*, but does not represent a distinct taxon. A number of additional but uncommon *Mucronella* species have been described from continental Europe, but have not yet been reported from Britain (see Corner, 1970; Maas Geesteranus, 1977).

Additional colour photographs may be found in Breitenbach & Kränzlin (1986: pl. 288), Moser & Jülich (1985: V *Mucronella* 1), and Ryman & Holmåsen (1984: 114).

Fig. 32. *Mucronella calva* (England, Surrey, Esher Common, N.W. Legon).

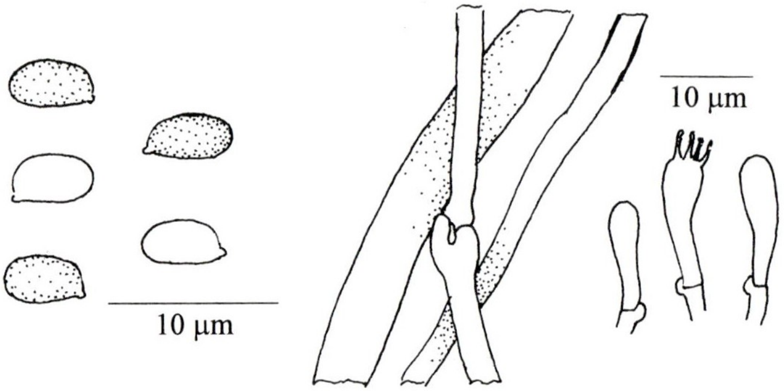

Fig. 33. *Mucronella calva* (England, Surrey, Esher Common, 19 Nov. 1995, P.J. Roberts, K(M) 32803) spores; generative-hyphae; basidia.

2. HERICIUM Pers.

in *Neues Mag. Bot.* 1: 109 (1794).

Friesites P. Karst. in *Medd. Soc. Fauna Fl. fenn.* 5: 41 (1879).
Creolophus P. Karst. in *Medd. Soc. Fauna Fl. fenn.* 5: 41 (1879).
Dryodon P. Karst. in *Rev. mycol.* 3(9): 19 (1881).

Fruitbody large, up to 30 cm across, branched or bracket-like, sometimes pileate, sessile or stalked. *Context* soft, fleshy. *Hymenophore* of conspicuous teeth or spines. *Spores* ellipsoid, smooth to punctate, hyaline, strongly amyloid. *Basidia* clavate, bearing 4 sterigmata. *Hyphal system* monomitic; *generative-hyphae* inflating in context, hyaline, amyloid or inamyloid, with clamp-connexions. *Gloeoplerous elements* present and conspicuous, yellowish, refractive, appearing in the hymenium as gloeocystidia. Lignicolous. Mainly north temperate. *Type species*: *Hericium coralloides* (Scop.) Pers.

Three *Hericium* species are known in the British Isles. A fourth species, *H. alpestre* Pers., occurs on conifers (*Abies*) in continental Europe.

The genera *Friesites* and *Dryodon* are obligate synonyms of *Hericium*, being based on the same type species. *Creolophus*, considered distinct by Donk (1962), appears indistinguishable from *Hericium*.

Key to British Species

1. Hyphae inamyloid, spores ellipsoid, smooth; fruitbody normally compact, with multiple pilei, often tiered .... 1. **Spine Face** (*H. cirrhatum*)
1. Hyphae inamyloid to strongly amyloid, spores broadly ellipsoid, finely punctate; fruitbody branched or compact, without distinct pilei **2**

 2. Fruitbody highly branched, spines up to 10 mm long .... 2. **Coral Spine Fungus** (*H. coralloides*)
 2. Fruitbody unbranched, spines up to 40 mm long ...... 3. **Monkey Head** (*H. erinaceum*)

1. SPINE FACE

Hericium cirrhatum (Pers.: Fr.) Nikol. in *Acta Inst. bot. Acad. Sci. URSS* Ser. II, fasc. 5: 343 (1950).
Hydnum cirrhatum Pers., *Syn. Meth. Fung.*: 558 (1801, as *H. cirratum*).
Hydnum diversidens Fr., *Syst. Myc.* I: 414 (1821).
Creolophus cirrhatus (Pers.: Fr.) P. Karst. in *Medd. Soc. Fauna Fl. fenn.* 5: 42 (1879).

Fruitbody up to 30 cm diam., irregular, broadly attached with no distinct stipe, typically divided into multiple, sometimes tiered, dimidiate pilei; pileal surface white to cream, occasionally flesh to salmon pink, becoming orange-brownish with age; uneven to warted or scrupose, sometimes with scattered sterile spines; whole fruitbody with strong fungal smell when dried. *Hymenophore* of conspicuous spines or teeth, 5–15 mm long, densely crowded, concolorous with pileus, becoming dark tan-brown when dried. *Context* soft, fleshy, unzoned, concolorous with pileus, becoming pale brown when dried. *Hyphal system* monomitic; *generative-hyphae* hyaline, inamyloid, acyanophilous, with clamp-connexions; in pileal context, most hyphae 6–16 μm diam., inflated and thick-walled, some narrow and thin-walled, with walls up to 3.5 μm thick; hyphae (excluding gloeoplerous elements) typically narrower in spines, 2–6(–12) μm diam., always thin-walled, and often with open, loop-like clamp-connexions. *Spores* 3.5–4 × 2.5 μm, ellipsoid, hyaline, strongly amyloid, smooth, with inconspicuous hilar appendix. *Basidia* 18–25 × 4–5 μm, narrowly clavate, bearing 4 sterigmata. *Gloeoplerous elements* abundant and conspicuous in the trama of the spines, sinuous and irregularly swollen up to 12 μm diam., with refractive, yellowish, oily contents; frequently emerging in the hymenium as gloeocystidia.

Habitat: on standing deciduous trees, sometimes high off the ground, and on fallen deciduous branches and logs. Most British specimens in K on beech (*Fagus*), but some also on ash (*Fraxinus*), elm (*Ulmus*), and oak (*Quercus*).

Distribution and frequency: widespread and locally common in southern England, particularly in native beech woods; much less frequent in the north, absent from Scotland, and apparently not recorded from Wales. Single specimen from Ireland (Co. Wexford) in K. Listed as 'vulnerable' in the provisional red data list of British fungi (Ing, 1992). June to November. Widespread throughout continental Europe. Reported from western United States by Harrison (1984) and from Canada by Ginns & Lefebvre (1993), but it is not clear that these collections are conspecific with European material.

Other remarks: probably the commonest of the larger hericioid fungi in Britain, *H. cirrhatum* can be distinguished from other *Hericium* species in the field by its possession of one or more distinct pilei. Microscopically the species is distinguished by its smooth spores, inamyloid hyphae, and thin-walled rather than thick-walled hyphae in the trama of the spines.

The pileate fruitbody and inamyloid hyphae led Donk (1962) to reintroduce Karsten's forgotten genus *Creolophus* as a 'tentative' generic name (Donk, 1962) to accommodate *Hericium cirrhatum*. Maas Geesteranus (1963) followed Donk, citing the thin-walled tramal hyphae as an additional distinction. Neither argument is compelling, and *Creolophus* is here treated as synonymous with *Hericium*.

Said to be edible, but of poor quality. Additional colour photographs can be found in Breitenbach & Kränzlin (1986: pl. 284), Dähncke (1993: 1039), Marchand (1976: pl. 337), Phillips (1981: 246), and Ryman & Holmåsen (1984: 112).

Fig. 34. *Hericium cirrhatum* (England, Hampshire, New Forest, Denny Wood, N.W. Legon).

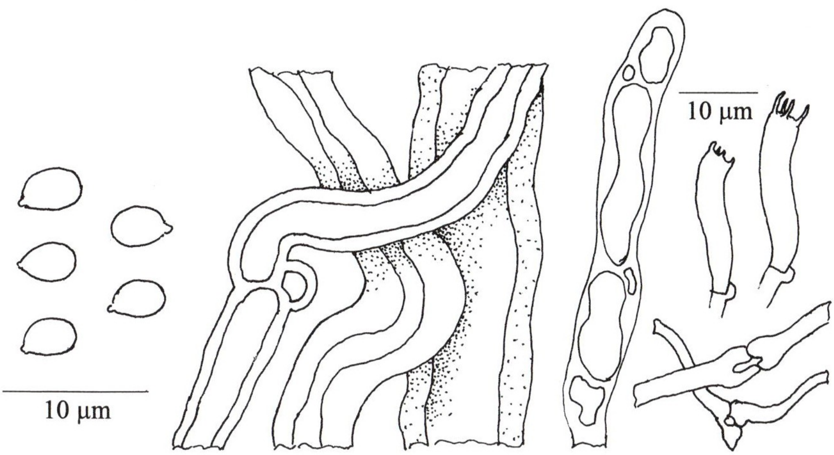

Fig. 35. *Hericium cirrhatum* (England, Hertfordshire, Nettleden, N. of Hemel Hempstead, 18 Aug., B. Young, K(M) 38426) spores; inflated, thick-walled hyphae from context; gloeoplerous hypha and narrow, clamped, generative hyphae; basidia.

2. CORAL SPINE FUNGUS

Hericium coralloides (Scop.: Fr.) Pers. in *Neues Mag. Bot.* 1: 109 (1794).
Hydnum coralloides Scop., *Fl. carniol.* 2: 472 (1772).
Hydnum clathroides Pallas, *Reise Prov. russ. Reich.* 2: 744 (1773).
Hydnum ramosum Bull., *Herb. France*: pl. 390 (1788).
Hericium clathroides (Pallas: Fr.) Pers., *Comm. Fung. clav.*: 23 (1797).
Hericium ramosum (Bull.) Let., *Hist. Descr. Champ.*: 43 (1826).

Fruitbody up to 25 cm across, apileate, branching extensively from a basal or lateral stipe to produce coral-like outgrowths of branchlets and spines. *Context* soft, white to pinkish becoming beige to yellowish brown with age. *Hymenophore* of conspicuous pendent spines or teeth, 5–10 mm long. *Spores* 3.5–5 × 3–4 µm, broadly ellipsoid, hyaline, strongly amyloid, finely punctate (at least when mounted in lactophenol cotton blue), hilar appendix inconspicuous. *Basidia* 16–35 × 3–5 µm, bearing 4 sterigmata. *Hyphal system* monomitic, *generative-hyphae* hyaline, varying within a single fruitbody from inamyloid to weakly or strongly amyloid, acyanophilous, with clamp-connexions; in stipe context, most hyphae inflated and thick-walled, some narrower and thin-walled, 3–25 µm diam., with walls up to 3.5 µm thick; most hyphae in spines narrower, 1.5–12 µm diam., some with slightly thickened refractive walls up to 1 µm diam. *Gloeoplerous elements* present and conspicuous in the trama of the spines, sinuous, often irregularly swollen and torulose, with yellowish, refractive contents, frequently emerging as gloeocystidia in the hymenium.

Habitat: on deciduous trunks and large logs. Most British specimens at K on beech (*Fagus*), but some also on ash (*Fraxinus*) and elm (*Ulmus*).

Distribution and frequency: uncommon and local in eastern England as far north as Nottinghamshire; apparently absent from the rest of the British Isles. Listed as 'vulnerable' in the provisional red data list of British fungi (Ing, 1992). September to November. Widespread but uncommon on the continent. Also known from North America (Ginns, 1985), where a number of additional species exist.

Other remarks: in the British Isles, *Hericium coralloides* can readily be identified by its attractive and conspicuous, branched, coral-like fruitbody. In the past, a large number of different growth forms were distinguished and named. However, Hallenberg (1983) has shown through culture studies that just two intersterile species exist in Europe, namely *Hericium coralloides* and *H. alpestre* Pers. The latter is only known from silver fir (*Abies*) and has not been recorded in the British Isles. Confusingly, Maas Geesteranus (1959) and many subsequent authors applied the name '*H. coralloides*' to *H. alpestre*, the names '*H. ramosum*' or '*H. clathroides*' being applied to *H. coralloides* as described here.

Hericium coralloides is normally described as having amyloid hyphae, but in many of the specimens examined at K this reaction is variable or absent. Such variability between specimens has previously been noted by Hallenberg (1983).

The species has also been called 'Cauliflower Spiny Cap' in Britain and 'Comb Tooth' in America. It is edible when young, but far too rare in Britain to be collected for food. Additional colour photographs can be found in Moser & Jülich (1985: V *Hericium* 1 & 3), Phillips (1981: 245, sub *H. ramosum*), and Ryman & Holmåsen (1984: 112). Note that the photographs in Breitenbach & Kränzlin (1986: pl. 286) and Marchand (1976: pl. 335) are of '*H. coralloides*' sensu Maas Geesteranus (1959) (= *H. alpestre*).

Fig. 36. *Hericium coralloides* (England, Hampshire, New Forest, Denny Wood, G. Dickson 656/84).

Fig. 37. *Hericium coralloides* (England, Hampshire, New Forest, Whitley Woods, G. Dickson 656/84).

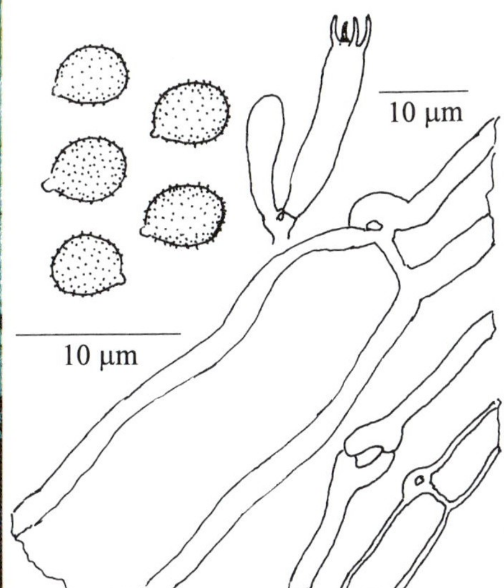

Fig. 38. *Hericium coralloides* (England, Essex, Epping Forest, 4 Oct. 1975, P.M. Kirk, K(M) 38534) spores; basidium; thick-walled, inflated, clamped hyphae in context, with thin-walled, uninflated, generative hyphae.

3. MONKEY HEAD

Hericium erinaceum (Bull.) Pers., *Comm. Fung. clav.* 27 (1797).
Hydnum erinaceum Bull., *Herb. France*: pl. 34 (1780).
Clavaria caputmedusae Bull., *Herb. France*: pl. 412 (1788).
Hericium hystrix Pers., *Comm. Fung. clav.* 27 (1797).
Steccherinum quercinum Gray, *Natural ang. Brit. Plants* I: 651 (1821).
Hericium unguiculatum Pers., *Mycol. europ.* 2: 153 (1825).
Hericium caputmedusae (Bull.) Pers., loc. cit.: 154 (1825).

Fruitbody sessile or short-stemmed, up to 25 cm across, normally comprising one or more large, apileate clusters of long, pendent spines. *Context* soft, fleshy, white becoming yellowish brown with age. *Hymenophore* of conspicuous teeth or spines, 10–40 mm long, white becoming sordid yellowish brown. *Spores* 5–6 × 4–5 µm, broadly ellipsoid, hyaline, strongly amyloid, finely but distinctly warted, hilar appendix inconspicuous. *Basidia* 25–35 × 6–7 µm, clavate, bearing 4 sterigmata. *Hyphal system* monomitic; *generative-hyphae* hyaline, mostly strongly amyloid, acyanophilous, with clamp-connexions; in the context, most hyphae are highly inflated and very thick-walled, often with just a narrow lumen remaining; a few are narrower, thin to thick-walled; 8.5–35 µm diam., walls up to 8.5 µm thick; narrower in spines, thin to thick-walled, 3–12 µm diam., walls up to 1 µm thick. *Gloeoplerous elements* abundant and conspicuous in the trama of the spines, sinuous, often irregularly swollen and torulose, with yellowish, refractive contents, frequently emerging as projecting gloeocystidia in the hymenium.

Habitat: on deciduous wood, mainly on old standing trees, often high off the ground.

Distribution and frequency: scattered but locally common in southern England, rare in the Midlands (one old specimen from Hereford in K), and apparently absent from the rest of the British Isles. Listed as 'vulnerable' in the provisional red data list of British fungi (Ing, 1992). August to November. Widespread but not common throughout continental Europe. Present in North America (Ginns, 1985).

Other remarks: distinguished from *H. coralloides* and *H. cirrhatum* by its non-pileate, non-branched, long-spined fruitbody, strongly amyloid hyphae, and comparatively large, distinctly punctate spores.

Hericium erinaceum has been given a host of additional names in English language books, including 'Hedgehog Fungus', 'Tree Hedgehog', 'Lion's Mane', 'Bear's Head', 'Bearded Tooth', 'Old Man's Beard', and others. The species is edible when young, but quickly becomes bitter and unpleasant. It is cultivated commercially in China, where it is valued as a traditional medicine for treating gastric complaints. In California it is cultivated and marketed as a gourmet food under the name 'Pom Poms' (Stamets, 1993).

Additional colour photographs may be found in Dähncke (1993: 1040), Marchand (1976: pl. 336), Phillips (1981: 245), and Ryman & Holmåsen (1984: 113).

Fig. 39. *Hericium erinaceum* (England, Hampshire, New Forest, Denny Wood, G. Dickson 656/84).

Fig. 40. *Hericium erinaceum* (England, Hampshire, New Forest, Frame Wood, G. Dickson 655/81).

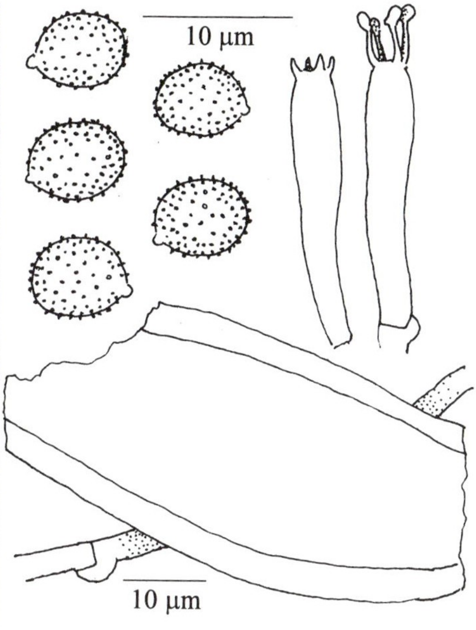

Fig. 41. *Hericium erinaceum* (England, Hampshire, New Forest, Denny Wood, 30 Oct.1994, N.W. Legon, K(M) 28087) spores; basidia; thick-walled, inflated hypha and thin-walled, clamped, generative hypha.

TOOTH-FUNGI

THELEPHORALES Corner ex Oberw.

in *Sydowia* 28: 361 (1976).

Fruitbody sessile or stalked, effused, pileate, or ramarioid. *Hymenophore* smooth, toothed, poroid, or lamellate. *Hyphal system* monomitic or dimitic, *generative hyphae* hyaline or coloured, inamyloid, with or without clamp-connexions. *Spores* warted to spiny, rarely smooth, hyaline or coloured, normally inamyloid. *Basidia* tubular to clavate, bearing (2–)4 sterigmata. *Cystidia* present in some taxa. Terricolous or lignicolous, often ectomycorrhizal. Worldwide. *Type family*: *Thelephoraceae* Chevall.

The *Thelephorales* are a diverse and possibly polyphyletic group of mainly ectomycorrhizal fungi characterized by their large and distinctly warted or spiny spores. Although it is comparatively easy to identify specimens of the *Thelephorales* to genus, identifying them to species often relies on highly variable macroscopic characters, including bright but unstable and ephemeral colours most of which are determined by derivatives of thelephoric acid (Stalpers, 1993). It is therefore essential to take detailed field notes when recording or collecting specimens. Many old herbarium specimens lacking field notes are effectively unidentifiable to species.

Following Donk (1961) and Maas Geesteranus (1975), two families are recognized within the *Thelephorales*, both of which occur in the British Isles. Only the stipitate hydnoid representatives are dealt with here. It should be noted that Jülich (1981) and Stalpers (1993) have adopted an alternative classification, placing all the stipitate hydnoid *Thelephorales* in the single family *Bankeraceae* Donk.

Other European members of the *Thelephorales* include *Tomentella* Pat. and its segregates, a large group of effused, patch-forming fungi often found under rotten wood or on damp ground; *Thelephora* Ehrh. ex Willd., a genus forming fan-like or coral-shaped fruitbodies, often around the roots of trees; and *Boletopsis* Fayod, a small genus of terrestrial, poroid fungi resembling boletes or 'ceps'.

Key to British Stipitate Hydnoid Families

1. Spore deposit whitish, spores hyaline; fruitbody often with strong spicy smell when dried .... 1. **Tooth-Fungi** (*Bankeraceae*) p. 59

1. Spore deposit coloured, spores coloured; no strong spicy smell when dried .... 2. **Corky Spine Fungi** & **Spiny Cap-Fungi** (*Thelephoraceae*) p. 74

6. BANKERACEAE Donk

in *Persoonia* 1: 405 (1961).

Fruitbody pileate and stipitate. *Context* fleshy, leathery, or woody, white or coloured, often with a spicy odour of fenugreek when dried. *Hymenophore* of teeth or spines. *Hyphal system* monomitic; *generative-hyphae* inamyloid, with or without clamp-connexions. *Spores* echinulate, hyaline, white in mass. Basidia tubular to clavate, bearing (2–)4 sterigmata. *Cystidia* absent or rare. *Gloeoplerous elements* often present, but not conspicuous. Terricolous, mycorrhizal. Temperate, less commonly tropical; worldwide. *Type genus: Bankera* Coker & Beers ex Pouzar.

Macroscopically the stipitate, hydnoid *Bankeraceae* closely resemble their counterparts in the *Thelephoraceae.* However, the spores of the *Bankeraceae* are always hyaline (white in mass), smaller, and more regular and even in outline than is typical for the stipitate, hydnoid *Thelephoraceae*. When dried, most members of the *Bankeraceae* also have a noticeable, spicy smell of fenugreek or curry powder, lacking in the stipitate hydnoid *Thelephoraceae.* Two genera occur in Europe.

Key to Genera

1. Hyphae in pileus context swollen up to 22 µm diam.; fruitbodies determinate (regularly shaped), not normally fused together .... **1. Drab** and **Violet Tooth-Fungi** (*Bankera*)

1. Hyphae in pileus not swollen, under 8 µm diam.; fruitbodies indeterminate (irregularly shaped), frequently fused together .... **2. Scented Spine Fungi** (*Phellodon*)

1. BANKERA Coker & Beers ex Pouzar

in *Ceská Mykol.* 9: 95 (1955).

Fruitbody pileate and stipitate, normally determinate (regularly shaped) and not fusing with other fruitbodies. Hymenophore of distinct teeth or spines. *Context* fleshy. *Hyphal system* monomitic; *generative-hyphae* hyaline, swollen in pileus up to 22 µm diam., inamyloid, without clamp-connexions. *Spores* ellipsoid, hyaline, inamyloid, echinulate. *Basidia* clavate, bearing 4 sterigmata. Terricolous, mycorrhizal with conifers. North temperate. *Type species: Bankera fuligineoalba* (J.C. Schmidt: Fr.) Pouzar. Two species in the British Isles.

Key to British Species

1. Associated with pine (*Pinus*); pileus typically with partly embedded needle debris, otherwise smooth .... 1. **Drab Tooth-Fungus** (*Bankera fuligineoalba*)

1. Associated with spruce (*Picea*); pileus typically without needle debris, breaking up into scales .... 2. **Violet Tooth-Fungus** (*Bankera violascens*)

1. DRAB TOOTH-FUNGUS

Bankera fuligineoalba (J.C. Schmidt: Fr.) Pouzar in *Ceská Mykol.* 9: 96 (1955).
Hydnum fuligineoalbum J.C. Schmidt, in Kunze & J.C. Schmidt, *Mykol. Hefte* 1: 88 (1817).
H. fragile Fr. in *Osvers. Svensk. Vet. Akad. Forh.* 8: 53 (1852).
Sarcodon fragilis (Fr.) P. Karst. in *Rev. Mycol.* 3(9): 20 (1881).
Tyrodon fuligineoalbum (J.C. Schmidt: Fr.) P. Karst. in *Bidr. Känn. Finl. Nat. Folk* 37: 91 (1882).
S. fuligineoalbum (J.C. Schmidt: Fr.) Quél., *Ench. Fung.* 189 (1886).
S. reticulatus Banker in *Mem. Torrey Bot. Club* 12: 139 (1905).
H. reticulatum (Banker) Sacc. & Trott., *Syll. Fung.* 21: 364 (1912).
H. virginianum Murr. in *Bull. Torrey Bot. Club* 67: 276 (1940).

Fruitbody pileate, 4–10(–15) cm diam., determinate (regularly shaped), with a short but distinct stipe 10–20 mm long; pileus convex becoming flat or depressed, surface smooth to squamulose, normally with embedded pine needle debris, not zoned, whitish becoming pale brown; stipe whitish below the spines, otherwise brown. *Hymenophore* of distinct teeth or spines up to 5 mm long, whitish at first becoming grey. *Context* of pileus up to 15 mm thick, white becoming brown; flesh in stipe grey-brown. *Hyphal system* monomitic; *generative-hyphae* mostly 3–6 µm diam., some swollen up to 12 µm diam. in pileus and stipe context, hyaline, inamyloid, without clamp-connexions. *Spores* 3.5–4 × 2.5–3 µm (ex 0.5 µm long spines), ellipsoid, hyaline, inamyloid, echinulate. *Basidia* 25–30 × 4.5–5 µm, tubular to weakly clavate, bearing 4 sterigmata.

Habitat: terricolous, mycorrhizal with *Pinus*, solitary or in troops among needle litter.

Frequency and distribution: uncommon in Scotland, mainly in native Caledonian pine forests; very rare and possibly extinct in England, with only a few old records from Berkshire (specimens in K), plus a more recent report from Yorkshire (Bramley, 1985). Listed as 'endangered' in the provisional red data list of British fungi (Ing, 1992). August to November. Widespread throughout continental Europe. Reported from North America (Baird, 1986b).

Other remarks: the habitat, embedded pine needle debris in the pileus, hyaline spores, and inflated pileal hyphae should make *Bankera fuligineoalba* easy to distinguish from other British stipitate hydnoid fungi.

Old British specimens in K under the name *Hydnum laevigatum* Fr. (a name of uncertain application) are *Bankera fuligineoalba*.

Additional colour photographs may be found in Breitenbach & Kränzlin (1986: pl. 272), Phillips (1981: p. 244), and Ryman & Holmåsen (1984: p. 106).

Fig. 42. *Bankera fuligineoalba* (Scotland, Inverness-shire, Nethybridge, Loch Garten, G. Dickson 823/85).

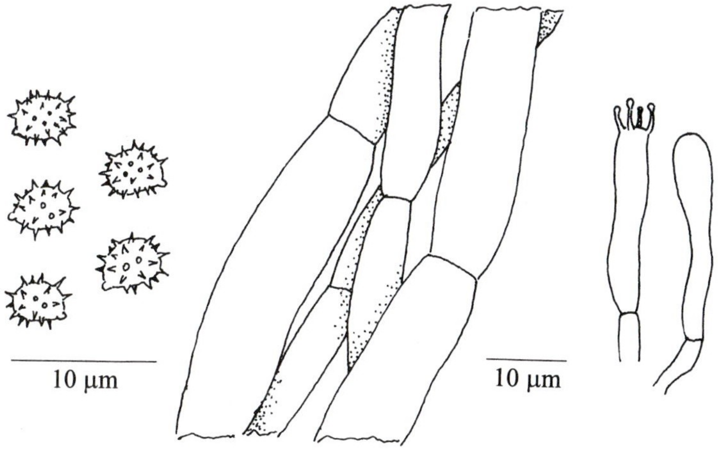

Fig. 43. *Bankera fuligineoalba* (Scotland, Perthshire, Black Wood of Rannoch, 4 Oct. 1967, P.D. Orton, K(M) 38375) spores; inflated hyphae in pileal context; basidia.

2. VIOLET TOOTH-FUNGUS

Bankera violascens (Alb. & Schwein.: Fr.) Pouzar in *Ceská Mykol.* 9: 96 (1955).
Hydnum violascens Alb. & Schwein., *Consp. Fung.*: 265 (1805).
Tyrodon violascens (Alb. & Schwein.) P. Karst. in *Bidr. Känn. Finl. Nat. Folk* 37: 91 (1882).
Sarcodon violascens (Alb. & Schwein) Quél. in *C. r. Ass. franç. Av. Sci.* 11: 399 (1883).

Fruitbody pileate, 5–10(–17) cm diam., determinate (regularly shaped), with a short but distinct stipe 10–25 mm long; pileus convex becoming flat or depressed, surface smooth at first, breaking up into conspicuous, adpressed scales; without embedded spruce needle debris, whitish becoming pale to purplish brown; stipe sometimes whitish below the spines, otherwise reddish to purplish brown. *Hymenophore* of distinct teeth or spines up to 5 mm long, whitish at first becoming grey. *Context* of pileus up to 15 mm thick, white becoming brown; flesh in stipe grey-brown. *Hyphal system* monomitic; *generative-hyphae* mostly 3–6 µm diam., some swollen up to 22 µm diam. in pileus and stipe context, hyaline, inamyloid, without clamp-connexions. *Spores* 3.5–4 × 3 µm (ex 0.5 µm long spines), ellipsoid, hyaline, inamyloid, echinulate. *Basidia* 25–30 × 4.5–5 µm, tubular to weakly clavate, bearing 4 sterigmata.

Habitat: terricolous, mycorrhizal with spruce (*Picea*), solitary or in troops among needle litter.

Frequency and distribution: unknown in the British Isles till 1996, when a specimen was collected in a spruce plantation in Scotland (Speyside); uncommon but widespread with native spruce in continental Europe.

Other remarks: the association with spruce and the needle-free, somewhat scaly pileus should be sufficient to distinguish *Bankera violascens* from *B. fuligineoalba*. Microscopically, the inflated hyphae appear to be even wider than in the latter species, and the spores tend towards the subglobose.

Though *B. fuligineoalba* is presumably native in Britain, *B. violascens* is clearly a recent introduction. Once established, it may be expected to spread into mature spruce plantations throughout the country.

Additional colour photographs may be found in Breitenbach & Kränzlin (1986: pl. 273) and Ryman & Holmåsen (1984: p. 106).

Fig. 44. *Bankera violascens* (Norway, Saaltdal, 27 Aug. 1993, P. Marstad).

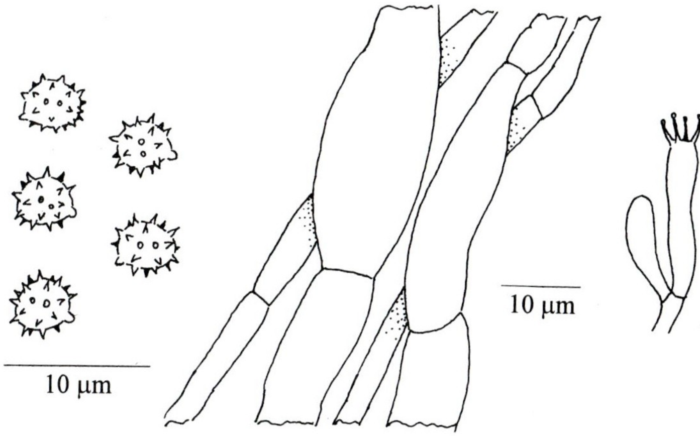

Fig. 45. *Bankera violascens* (Scotland, Loch Morlich, 22 July 1996, E. Farquharson, E) spores; inflated hyphae in pileal context; basidia.

Fig. 46. *Bankera fuligineoalba* (Scotland, 1879, as '*Hydnum fragile*'), from an original painting by M.C. Cooke.

2. PHELLODON P. Karst.

in *Revue mycol.* 3(9): 19 (1881).

Fruitbody pileate and stipitate, normally indeterminate (irregularly shaped), often fusing with other fruitbodies to form compound growths; pileus flat to depressed, occasionally infundibuliform, normally tomentose at first, then becoming pitted, ridged, or corrugated, often in zones, sometimes producing outgrowths or small secondary pilei. *Hymenophore* of teeth or spines. *Context* in all European species with odour of fenugreek when dried. *Hyphal system* monomitic; *generative-hyphae* inamyloid, less than 6 µm diam., without clamp-connexions in all European species. *Spores* ellipsoid, hyaline, inamyloid, echinulate. *Basidia* tubular to weakly clavate, bearing 4 sterigmata. *Cystidia* present in some species as occasional poorly differentiated cystidioles. Terricolous, mycorrhizal. Worldwide. *Type species: Phellodon niger* (Fr.) P. Karst.

The genus can be distinguished from other stipitate hydnoid fungi by its hyaline spores, narrow hyphae, tendency to form irregular, compound fruitbodies, and frequently a spicy smell of fenugreek or curry powder when dried. The irregular, indeterminate growth and variability of colours make identifying collections to species rather difficult, and it is essential to make notes on colour, habitat, and surface features at the time of collection. The reaction of the context and hyphae to alkaline solutions (KOH or ammonia) is variable, even within a single fruitbody, but normally strong in *P. niger*. There are four species in the British Isles and Europe. None are edible.

Key to Species

1. Stipe surface smooth, polished **2**
1. Stipe surface tomentose **3**

2. Pileus flesh greyish 1. **Black & White Scented Spine Fungus** (*Phellodon melaleucus*)
2. Pileus flesh pallid or yellowish brown
.... 2. **Goblet Scented Spine Fungus** (*Phellodon tomentosus*)

3. Pileus and stipe becoming dark grey to black; stipe context blackish, woody, markedly staining green in KOH or ammonia solution
.... 3. **Black Scented Spine Fungus** (*Phellodon niger*)
3. Pileus and stipe normally brownish; stipe context pale brown to grey-brown, not woody, not or barely staining green in KOH or ammonia solution **4**

4. Pileus surface concentrically zoned, tan to brown, normally growing with conifers
.... **Goblet Scented Spine Fungus** (*Phellodon tomentosus*)
4. Pileus surface not concentrically zoned, normally growing with deciduous trees
.... 4. **Strongly Scented Spine Fungus** (*Phellodon confluens*)

1. BLACK & WHITE SCENTED SPINE FUNGUS

Phellodon melaleucus (Schwein.: Fr.) P. Karst. in *Revue mycol.* 3(9): 19 (1881).
Hydnum melaleucum Schwein. in Fr., *Obs. mycol.* 1: 141 (1815).
H. leptopus var. *graveolens* Pers., *Mycol. Eur.* 2: 171 (1825).
H. delicatum Schwein., *Syn. N. Am. Fung.*: 161 (1834).
H. graveolens (Pers.) Fr., *Epicr. Syst. Myc.*: 509 (1838).
Hydnellum graveolens (Pers.) P. Karst. in *Medd. Soc. Fauna Fl. fenn.* 5: 41 (1879).
Hydnellum melaleucum (Schwein.: Fr.) P. Karst., loc. cit.: 41.
P. graveolens (Pers.: Fr.) P. Karst. in *Bidr. Kann. Finl. Nat. Folk.* 37: 96 (1882).
Calodon melaleucus (Schwein.: Fr.) Quélet, *Ench. Fung.*: 191 (1886).
C. graveolens (Pers.) Quélet, loc. cit..: 191
P. ellisianus Banker, *Mem. Torr. Bot. Club* 12: 174 (1906).
P. brunneoroseus Snell & al., *Lloydia* 19: 171 (1956).

Fruitbody stipitate and pileate, normally fusing to form compound multipileate growths; individual pilei up to 6 cm diam., flat to depressed, tomentose at first becoming wrinkled or ridged in concentric zones; white at first becoming greyish, reddish, olivaceous, or violaceous brown, dark brown or blackish; *stipe* up to 4 cm long, usually smooth to fibrillose, not tomentose, similar colour to pileus. *Hymenophore* of teeth or spines up to 2 mm long, white becoming grey to brown. *Context* fleshy, up to 2 mm thick in the pileus, brown to greyish brown, odour of fenugreek when dried. *Spores* 3.5–5 × 3–4.5 µm (ex 0.5 µm long spines), subglobose, hyaline, inamyloid. *Basidia* 20–30 × 5–6 µm, clavate, bearing 4 sterigmata. *Hyphal system* monomitic; *generative-hyphae* 3–7 µm diam., uninflated, pale brown, inamyloid, clamp-connexions absent.

Habitat: terricolous, in old woodland, mycorrhizal with deciduous and coniferous trees. British specimens reported with oak (*Quercus*), birch (*Betula*), beech (*Fagus*), chestnut (*Castanea*), spruce (*Picea*), and pine (*Pinus*).

Distribution and frequency: widespread and rather common in England, Scotland, Wales, and Eire. Reported as 'very rare' in Northern Ireland by Muskett & Malone (1980). Listed as 'vulnerable' in the provisional red data list of British fungi (Ing, 1992). August to November. Widespread throughout continental Europe. Reported from North America (Baird, 1986b).

Other remarks: distinguished from other *Phellodon* species by its smooth stipe, zoned pileus, and generally grey or darkish context.

Additional colour photographs may be found in Breitenbach & Kränzlin (1986: pl. 269), Phillips (1981: 245), and Ryman & Holmåsen (1984: 105).

Fig. 47. *Phellodon melaleucus* (Scotland, Inverness-shire, Nethybridge, Abernethy Reserve, G. Dickson 823/85).

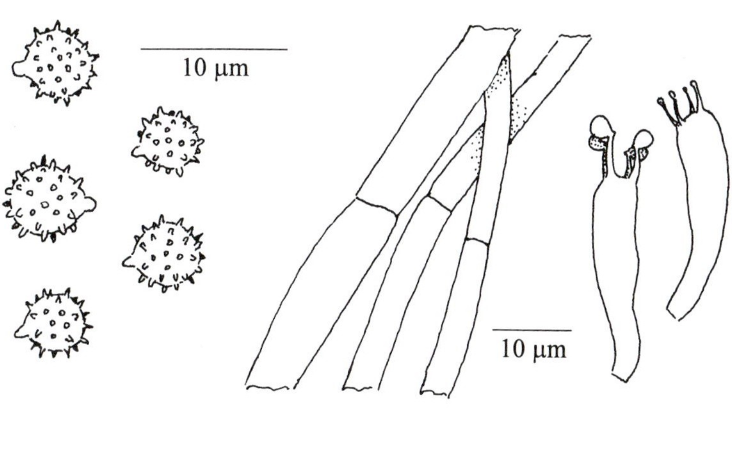

Fig. 48. *Phellodon melaleucus* (England, Devon, Slapton Wood, 6 Oct. 1974, D.L. Hawksworth, K(M) 38809) spores; uninflated hyphae in context; basidia.

2. GOBLET SCENTED SPINE FUNGUS

Phellodon tomentosus (L.: Fr.) Banker in *Mem. Torrey bot. Club* 12: 171 (1906).
Hydnum tomentosus L., *Sp. Pl.* 2: 1178 (1753).
H. cyathiforme Schaeff., *Fung. Icon.* 4: 93 (1774).
H. leptopus Pers., *Mycol. Eur.* 2: 170 (1825).
H. coriaceo-membranaceum Schwein. in *Trans. Am. Phil. Soc.* 4: 162 (1834).
Hydnellum cyathiforme (Schaeff.) P. Karst. in *Medd. Soc. fauna Fl. fenn.* 6: 15 (1879).
Phellodon cyathiforme (Schaeff.) P. Karst. in *Revue mycol.* 3(9): 19 (1881).
Calodon cyathiformis (Schaeff.) Quélet, *Ench. Fung.*: 191 (1886).
P. coriaceo-membranaceum (Schwein.) Banker in *Mem. Torrey Bot. Club* 12: 172 (1906).
C. tomentosus (L.: Fr.) Maire in *Publ. Inst. Bot. Barcelona* 3(4): 36 (1937).

Fruitbody stipitate and pileate, normally fusing to form compound multipileate growths; individual pilei up to 4 cm diam., typically depressed, thinly tomentose at first then becoming wrinkled or concentrically zoned; white at first becoming tan to brown; *stipe* up to 4 cm long, usually tomentose, coloured as pileus. *Hymenophore* of teeth or spines up to 2 mm long, white becoming greyish-pink. *Context* fleshy, up to 2 mm thick in the pileus, pallid to yellowish brown, odour of fenugreek when dried. *Spores* 3–4 × 3–4 µm (ex 0.5 µm long spines), globose to subglobose, hyaline, inamyloid. *Basidia* 18–30 × 4–6 µm, clavate, bearing 4 sterigmata. *Hyphal system* monomitic; *generative-hyphae*, 2.5–6 µm diam., uninflated, hyaline to pale brownish, inamyloid, clamp-connexions absent.

Habitat: terricolous, in old woodland, mycorrhizal with coniferous trees. British specimens reported with pine (*Pinus*).

Distribution & frequency: widespread but uncommon in England and Scotland; also reported from Eire (Muskett & Malone, 1980). Listed as 'endangered' in the provisional red data list of British fungi (Ing, 1992). August to November. Widespread throughout continental Europe. Reported from North America (Baird, 1986b).

Other remarks: distinguished by its zoned, normally brownish pileus, pale brownish context, and occurrence with conifers. Additional colour photographs may be found in Breitenbach & Kränzlin (1986: pl. 271), Marchand (1976: pl. 340), Phillips (1981: 245), and Ryman & Holmåsen (1984: 105).

Fig. 49. *Phellodon tomentosus* (Sweden, Sundsvall, 22 Aug. 1992, P. Marstad).

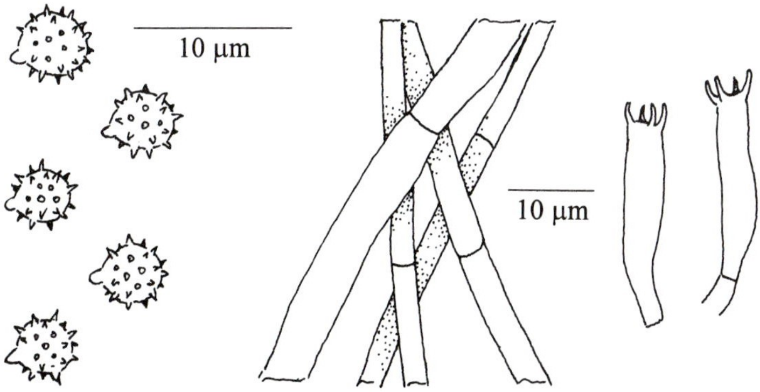

Fig. 50. *Phellodon tomentosus* (Scotland, Inverness-shire, Abernethy, Loch Garten, 11 Nov. 1991, T. Læssøe, K(M) 17850) spores; uninflated hyphae in context; basidia.

3. BLACK SCENTED SPINE FUNGUS

Phellodon niger (Fr.) P. Karst. in *Revue mycol.* 3(9): 19 (1881).
Hydnum nigrum Fr., *Obs. mycol.* 1: 134 (1815).
H. olidum Berk. in *Journ. Linn. Soc. Bot.* 16: 51 (1877).
Hydnellum nigrum (Fr.) P. Karst. in *Medd. Soc. Fauna Fl. fenn.* 5: 41 (1879).
Calodon niger (Fr.) Quélet, *Ench. Fung.*: 191 (1886).
Hydnum cuneatum Lloyd, *Myc. Writ.* 7: 1345 (1925).

Fruitbody stipitate and pileate, normally fusing to form compound multipileate growths; individual pilei up to 5 cm diam., flat to depressed or infundibuliform, tomentose at first becoming pitted, wrinkled, or ridged in concentric zones; whitish at first, sometimes with violaceous tints, then darkening from centre to grey, grey-brown, or blackish; *stipe* up to 4 cm long, fibrillose to tomentose, concolorous with pileus. *Hymenophore* of teeth or spines up to 4 mm long, grey. *Context* fleshy, up to 5 mm thick in the pileus, dark grey to black, turning blue-green in KOH or ammonia solution, odour of fenugreek when dried. *Spores* 3.5–5 × 3–4 µm (ex 0.5 µm long spines), broadly ellipsoid, hyaline, inamyloid. Basidia clavate, 25–40 × 5–7 µm, bearing 4 sterigmata. *Hyphal system* monomitic; *generative-hyphae* 2.5–5 µm diam., uninflated, dark brown to blackish, encrusted with fine bluish-green granules in KOH or ammonia solution, inamyloid, clamp-connexions absent.

Habitat: terricolous, in old woodland, mycorrhizal with deciduous and coniferous trees. British specimens reported with chestnut (*Castanea*), beech (*Fagus*), and pine (*Pinus*).

Distribution & frequency: widespread but uncommon in England, Scotland, and Wales; also reported from Eire (Muskett & Malone, 1980). Listed as 'rare' in the provisional red data list of British fungi (Ing, 1992). August to November. Widespread throughout continental Europe. Reported from North America (Baird, 1986b).

Other remarks: distinguished by its dark grey to black colours, similarly coloured context, and the marked greening of the hyphae in KOH or ammonia solution. Additional colour photographs may be found in Breitenbach & Kränzlin (1986: pl. 270), Dähncke (1993: 1046), Marchand (1976: pl. 339), Moser & Jülich (1985: V *Phellodon* 1), Phillips (1981: 245), and Ryman & Holmåsen (1984: 104).

Fig. 51. *Phellodon niger* (England, Hampshire, New Forest, Norleywood, G. Dickson 807/85).

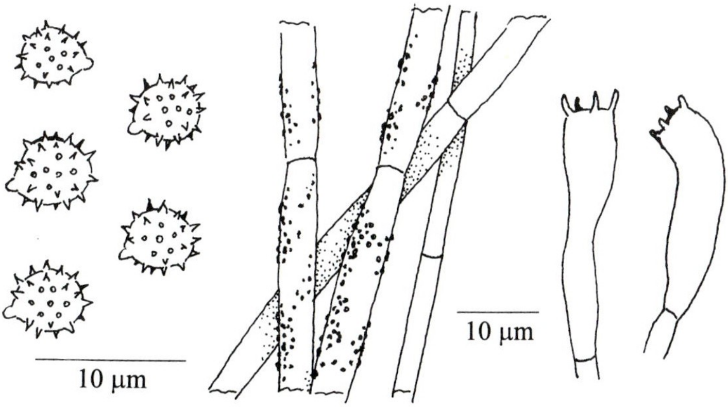

Fig. 52. *Phellodon niger* (England, Surrey, Windsor Great Park, 19 Oct. 1985, E.E. Green, K(M) 38663) spores; uninflated, partly encrusted hyphae in context; basidia.

4. STRONGLY SCENTED SPINE FUNGUS

Phellodon confluens (Pers.) Pouzar in *Ceská Mykol.* 10: 74 (1956).
Hydnum confluens Pers., *Mycol. europ.* 2: 165 (1825).
H. amicum Quél. in *Grevillea* 8: 115 (1880).
Sarcodon amicus (Quél.) Quél., *Ench. Fung.*: 189 (1886).
Hydnellum vellereum Peck in *Ann. Rep. New York St. Mus.* 50: 110 (1897).
P. vellereum (Peck) Banker in *Mem. Torrey Bot. Club* 12: 168 (1906).
P. amicus (Quél.) Banker in *Mycologia* 5: 62 (1913).
Hydnellum amicum (Quél.) Ragab in *Mycologia* 45: 944 (1953).

Fruitbody stipitate and pileate, normally fusing to form compound multipileate growths; individual pilei up to 6 cm diam., flat to depressed, tomentose at first, becoming roughened, pitted, and irregular at the centre, not or scarcely zoned, white at first becoming variously pale ochre to buff, grey, or brownish. *Stipe* indistinct, up to 4 cm long, tomentose (especially towards the base), white at first becoming greyish brown. *Hymenophore* of teeth or spines up to 2 mm long, white becoming greyish to brown, sometimes tinged violaceous. *Context* fleshy, up to 5 mm thick in the pileus, white becoming greyish to brownish, darker in the stipe, odour of fenugreek when dried. *Spores* 4.5 × 3–4 μm (ex 0.5 μm long spines), subglobose, hyaline, inamyloid. *Basidia* 20–35 × 5–7 μm, clavate, bearing 4 sterigmata. *Hyphal system* monomitic; *generative-hyphae* 2.5–5 μm diam., uninflated, pale brown, inamyloid, clamp-connexions absent.

Habitat: terricolous in old woodland, mycorrhizal with deciduous trees. British specimens reported with beech (*Fagus*), chestnut (*Castanea*), oak (*Quercus*), and birch (*Betula*). Some overseas collections also reported with conifers.

Distribution & frequency: widespread but not common in southern England, with isolated records from Scotland. Not reported from Ireland. Listed as 'endangered' in the provisional red data list of British fungi (Ing, 1992). August to November. Widespread throughout continental Europe. Reported from the eastern United States (Baird, 1986b).

Other remarks: distinguished from other *Phellodon* species by its typically brown colouration, tomentose stipe, lack of conspicuous zoning on the pileus, and occurrence with deciduous trees. Additional colour photographs may be found in Breitenbach & Kränzlin (1986: pl. 268), Marchand (1976: pl. 338), and Phillips (1981: 245).

Fig. 53. *Phellodon confluens* (Scotland, Inverness-shire, Nethybridge, Abernethy Reserve, G. Dickson 406/78).

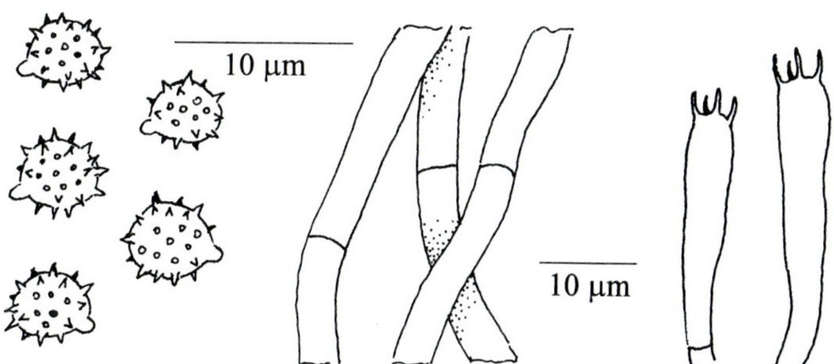

Fig. 54. *Phellodon confluens* (England, Berkshire, Swinley, near Ascot, 23 Oct. 1982, E.E. Green, K(M) 38701) spores; uninflated hyphae in context; basidia.

7. THELEPHORACEAE Chevall.

Fl. gen. env. Paris: 84 (1826).

Fruitbody terricolous or lignicolous, annual, variable in form, commonly resupinate, effuso-reflexed, clavarioid or pileate, stipitate or not, commonly ectomycorrhizal. *Hymenophore* smooth or warted, hydnoid, rarely poroid. *Context* coriaceous or fleshy. *Hyphal system* monomitic, less commonly dimitic; *generative-hyphae* thin- or sometimes thick-walled, sometimes inflating, hyaline to brown, clamp-connexions present or absent, non-amyloid. *Gloeoplerous hyphae* sometimes present. *Cystidia* usually absent. *Basidia* clavate to cylindric-clavate, (2–)4-sterigmate. *Spores* hyaline or pigmented, warted to echinulate. *Type genus*: *Thelephora* Ehrh. ex Willd.

The family is worldwide in distribution. Pileate, stipitate species are ranged in several genera which, although usually referred to this family, are placed by Stalpers (1993) in *Bankeraceae*. They have a hydnoid hymenophore, with the exception of *Boletopsis*, which is poroid.

Key to British Hydnoid Genera

1. Context corky or woody, zoned; generative-hyphae not inflating; growth irregular .... **1. Hydnellum**

1. Context brittle, not zoned; generative-hyphae inflating; growth regular .... **2. Sarcodon**

1. HYDNELLUM P. Karst.

in *Medd. Soc. Fauna Fl. fenn.* 5: 41 (1879)

Fruitbody annual, terrestrial, pileate, stipitate, simple or concrescent. *Pileus* velutinous or felted to somewhat scaly or scrobiculate, whitish, yellow, brown to purple-brown or orange-brown or blue-grey, sometimes with distinct radial grooves or ribs and commonly with concentric zonation. *Stipe* usually central, solid, tomentose or velutinous to spongy, concolorous with pileus or not. *Hymenium* covering spines on the underside of the pileus. *Spines* slender, subulate, mostly purplish-brown at maturity, usually decurrent. *Context* fibrillose, corky to woody, variously pigmented and zoned, in some species yielding a deep green or blue-green stain in alkali solution. *Odour* various but never of fenugreek; *taste* mild or acrid. *Hyphal system* monomitic; *generative-hyphae* thin-walled, not inflating, septate, clamp-connexions present or not; *gloeoplerous hyphae* present or not. *Spore deposit* brown. *Spores* tuberculate to warted, subglobose or broadly ellipsoid, irregular in outline. *Basidia* typically narrowly clavate, usually 4-spored. *Cystidia* lacking. *Type species*: *Hydnellum suaveolens* Scop.: Fr.

The genus is widely distributed in the northern hemisphere, occuring throughout Europe and North America, and is also present in the Eastern tropics. Sixteen species are recognized in Europe by Maas Geesteranus (1975), many of them now rare or decreasing and considered as endangered or even extinct in some areas. Identification of species commonly presents difficulties in the absence of fresh specimens; although presence or absence of clamp-

connexions is of specific importance, microscopic characters are otherwise similar in many cases. It is essential, therefore, that full, detailed field notes are made for all collections of *Hydnellum*. Colour of the flesh offers particularly important specific characters, and taste and odour are also of value in distinguishing species. Habitat may also offer helpful characters as species are ectomycorrhizal, although some occur with both deciduous and coniferous trees.

Seven species are recognized in the present account, but most of them are rare and considered as vulnerable or endangered. *Hydnellum compactum* (Pers.: Fr.) P. Karst. and *H. mirabile* (Fr.) P. Karst. have also been reported as British, but this appears to be on the basis of misidentifications or misapplication of names, and their presence here has not been confirmed in the present study. However, the genus is not well known in Britain and it is possible that species additional to those included in the present account await discovery here.

Key to Species

1. Clamp-connexions present on context hyphae **2**
1. Clamp-connexions lacking on context hyphae **3**

2. Flesh with blue colours present, at least in pileus; pileus blue-grey at least at centre, not exuding blood-red droplets; taste mild .. **1. Blue Corky Spine Fungus** (*H. caeruleum*)
2. Flesh lacking blue colours; pileus reddish or red-brown at least at centre, often exuding blood-red droplets when fresh; taste acrid **2. Brown Corky Spine Fungus** (*H. peckii*)

3. Pileus distinctly orange or orange-brown, at least at centre; context with distinct orange tints, especially in stipe .... **3. Orange Corky Spine Fungus** (*H. aurantiacum*)
3. Pileus some shade of brown, sometimes pinkish- or red-brown, not orange; context lacking orange tints **4**

4. Pileus with distinct radial grooves and ridges and commonly with concentric zonation; usually depressed or infundibuliform, and often scrobiculate **5**
4. Pileus without obvious radial structures except sometimes at margin with age, concentric zonation present or not; convex or depressed, velutinous or matted **6**

5. Context and spines rapidly exuding a dark green to blue-green stain in alkali solution; concentric zonation well developed; spores smaller, to 5.5×3.5 µm .... **4. Concrescent Corky Spine Fungus** (*H. concrescens*)
5. Context and spines not exuding green stain in alkali, with brownish stain or none; concentric zonation less well developed; spores larger, to 6.5×5.5 µm .... **5. Pitted Corky Spine Fungus** (*H. scrobiculatum*)

6. Growing with conifers; context paler, pinkish-brown .... **6. Reddish Brown Corky Spine Fungus** (*H. ferrugineum*)
6. Growing with *Fagaceae*; flesh darker, brown to red-brown .... **7. Velvety Corky Spine Fungus** (*H. spongiosipes*)

1. BLUE CORKY SPINE FUNGUS

Hydnellum caeruleum (Hornem.) P. Karst. in *Medd. Soc. Fauna Flora fenn.* 5: 41 (1879).
Hydnum caeruleum Hornem., *Fl. dan.* 8 (fasc. 23): 7 (1808).
Hydnum suaveolens ssp. *caeruleum* (Hornem.) Fr., *Epicr. Syst. Mycol.*: 507 (1838)
Calodon caeruleus (Hornem.) P. Karst. in *Bidr. Känn. Finl. Nat. Folk* 37: 106 (1882)
Phellodon caeruleus (Hornem.) R. Nav. in *Natuurwet. Tijdschr., Antw.* 5: 67 (1923)

Fruitbody solitary or concrescent. *Pileus* 2–7(–11) cm across, convex or plane to depressed, scrobiculate, velutinous at first becoming glabrous with age, often with concentric marks or grooves, rarely with radial markings, with bluish tinge at first, or blue-grey, often becoming brown with age; margin usually paler, whitish. *Stipe* 1.5–5 cm long, 1–2.5 cm thick, cylindric to ventricose, velutinous, orange to orange-brown, binding debris. *Spines* decurrent, 3–6 mm long, bluish at first, later brownish to purple-brown. *Context* whitish to pale orange, zoned blue-grey in pileus, more deeply orange-brown in stipe; *odour* mealy; *taste* mild. *Context hyphae* 4–8.5 µm diam., thin-walled, simple or branched, with clamp-connexions at some septa, hyaline or with blue or yellowish pigment; *hyphae in spines* 3–4.5 µm diam., hyaline to pale yellowish, sparsely septate, simple or branched. *Spores* 4.5–6 × 4–4.5 µm, pale yellow to almost hyaline, ellipsoid to irregular in outline, tuberculate. *Basidia* 27–35 × 6–7 µm, narrowly clavate, 4-sterigmate.

Habitat: usually in coniferous woodland, with pine (*Pinus*) or spruce (*Picea*), on acid, sandy soils. Occasionally with beech (*Fagus*).

Distribution and frequency: found throughout Europe and also present in North America and parts of Asia, but regarded as uncommon or rare. In Britain this species is evidently very rare, being known hitherto only from native pine woods at a few localities in the Scottish Highlands.

Other remarks: distinguished by the blue-zoned flesh of the pileus, presence of scattered clamp-connexions, and mealy odour. It is similar to *H. suaveolens*, not yet recorded from Britain, which differs in its strong odour of anise and distinct blue colours in the stipe flesh.

Many records of *Hydnellum compactum* (Pers.: Fr.) P. Karst., not confirmed as British in the present study, were due to misidentification of *H. caeruleum*. *Hydnellum compactum* has pale flesh, and lacks clamp-connexions. Furthermore, although commonly with olivaceous tints in the pileus, it lacks blue colours, and occurs with *Fagaceae*, especially *Quercus* (Maas Geesteranus 1975; Stalpers 1993), never with conifers.

Additional coloured illustrations of *H. caeruleum* can be found in Breitenbach & Kränzlin (1986: pl. 259), Cetto (1979–84: 761, 1987a: 237), Marchand (1976: 342), Moser & Jülich (1985: V), Phillips (1981: 242), and Ryman & Holmåsen (1984: 97).

Fig. 55. *Hydnellum caeruleum* (Scotland, Inverness-shire, Abernethy Reserve, Nethybridge, G. Dickson 477).

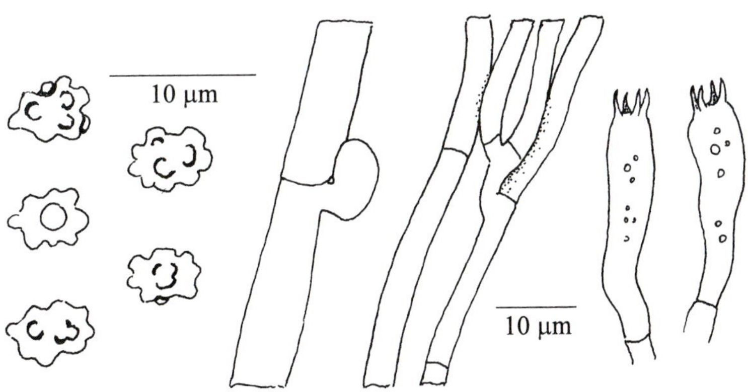

Fig. 56. *Hydnellum caeruleum* (Scotland, Strathclyde, Inveraray Woods, 28 Sep. 1986, A. Outen, K(M) 40696) spores; clamped hyphae in context, unclamped hyphae in hymenium; basidia.

2. BROWN CORKY SPINE FUNGUS

Hydnellum peckii Banker in Peck, *Bull. N.Y. St. Mus.* 157: 28 (1912).
Hydnum peckii (Banker in Peck) Trott. in Syll. Fung. 23: 470 (1925).
Calodon peckii (Banker in Peck) Snell & Dick in *Lloydia* 19: 163 (1956).
Hydnellum diabolus Banker in *Mycologia* 5: 194 (1913) teste Maas Geesteranus (1969; 1975).
Hydnum diabolus (Banker) Trott. in *Syll. Fung.* 23: 470 (1925).
Calodon diabolus (Banker) Murrill in *Florida agr. Exp. Stn Bull.* No. 478: 11 (1951).

Fruitbody usually gregarious, solitary to caespitose or concrescent. *Pileus* 3–8(–12) cm across, irregularly convex or pulvinate, becoming depressed with age, velutinous at first, becoming matted, scrobiculate, whitish when young, with age becoming red-brown to dark brown from centre outwards, exuding blood-red droplets in damp conditions, radially grooved. *Stipe* usually short, 2–4 cm long, 5–20 mm thick, cylindric or tapered, surface at least in lower part thickly velutinous, matted, concolorous with pileus. *Spines* decurrent, 3–4.5 mm long, slender, whitish at first, becoming vinaceous. *Context* pale pinkish-brown, with darker zones, sometimes with yellowish spots, darker in stipe; *odour* slight; *taste* hot, acrid. *Context hyphae* hyaline to pale brownish, thin-walled, septate, with frequent clamp-connexions; *hyphae in spines* simple or branched, sometimes flexuous, thin-walled, hyaline to pale brown, septate. *Spores* (4.5–)5–6(–6.5) × 3.5–5 μm, yellow-brown, broadly ellipsoid to subglobose, tuberculate. *Basidia* 28–34 × 5–6.5 μm, cylindric-clavate, without basal clamp-connexion, (2–)4-spored.

Habitat: coniferous woods, with pine (*Pinus*) and spruce (*Picea*), on acid soils.

Distribution and frequency: present throughout Europe, though rare and considered endangered in some areas; also known from north-eastern and western parts of North America. In Britain this species is rare and apparently confined to native pine woods in the Scottish Highlands.

Other remarks: the species is characteristic and may be readily distinguished by the acrid flesh, presence of clamp-connexions, reddish-brown pileus and the characteristic red droplets exuded when the fungus is actively growing. *Hydnellum ferrugineum* may also exude red droplets from actively developing fruitbodies, but has a mild taste and lacks clamp-connexions.

The synonymy with this species of *H. diabolus* as concluded by Maas Geesteranus (1969) was not supported by Baird (1986a,b). Other authors, such as Coker & Beers (1951), have also recognized the species as distinct.

Additional coloured illustrations of *H. peckii* can be found in Breitenbach & Kränzlin (1986: pl. 264), Cetto (1987a: 227), Marchand (1976: 345), Phillips (1981: 243), and Ryman & Holmåsen (1984: 99).

Fig. 57. *Hydnellum peckii* (Scotland, Inverness-shire, Coylum Bridge, G. Dickson 800/83).

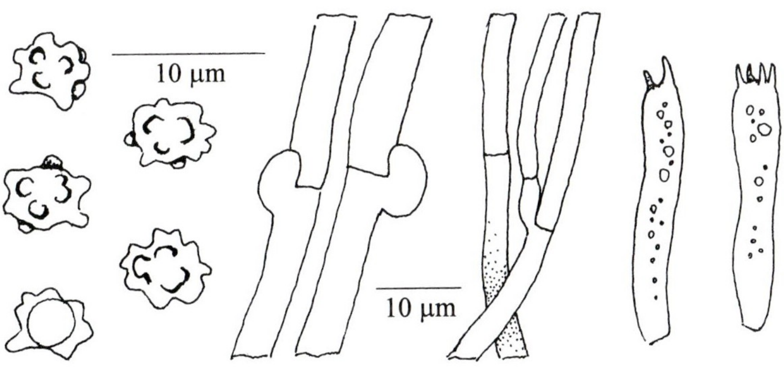

Fig. 58. *Hydnellum peckii* (Scotland, Inverness-shire, Culbin Forest, 25 Aug. 1993, J. Weightman, K(M) 25142) spores; clamped hyphae in context; unclamped hymenial hyphae; basidia.

3. ORANGE CORKY SPINE FUNGUS

Hydnellum aurantiacum (Alb. & Schwein.: Fr.) P. Karst. in *Medd. Soc. Fauna Flora fenn.* 5: 41 (1879).
Hydnum suberosum var. *aurantiacum* Batsch, *Elenchus Fung.* cont. sec.: 103 (1789).
Hydnum aurantiacum (Batsch: Fr.) Alb. & Schwein., *Conspectus Fung.*: 265 (1805); Fr., *Syst. mycol.* 1: 403 (1821).
Calodon aurantiacus (Batsch: Fr.) P. Karst. in *Rev. Mycol.* 3: 20 (1881).
Phaeodon aurantiacus (Batsch: Fr.) J. Schroet. in *Kryptog. Fl. Schles.* 3(1): 459 (1888).

Fruitbody solitary or concrescent. *Pileus* 2–9 cm across, often convex when young, becoming plane to depressed, surface often with irregular excrescences, radially fibrillose, ridged or wrinkled, tomentose at first, becoming smooth, pale orange-brown, darker at centre, sometimes darker brown with age, margin usually paler, whitish to pale orange-yellow, concentric zonation weak. *Stipe* 2–5 × 1–2.5 cm, cylindric or tapered downwards, often swollen at the base, sometimes branched, orange to orange-brown, velutinous to tomentose. *Spines* decurrent, 2–4 mm long, slender, whitish at first, becoming brown or orange-brown. *Context* whitish or pale yellow-orange in pileus, darker orange below, often zoned with darker lines. *Odour* slight; *taste* mild. *Context hyphae* 4–5.5 µm diam., almost hyaline, thin-walled, often branched, remotely septate, lacking clamp-connexions; *hyphae in spines* slender, 3–5 µm diam., hyaline to pale yellowish, sometimes branched, thin-walled. *Spores* 5.5–7 × 4–5 µm, pale yellowish, tuberculate. *Basidia* 35–50 × 6–8 µm, narrowly clavate, (2–)4-sterigmate.

Habitat: known in Europe from coniferous and mixed woodland, mostly in association with pine (*Pinus*) on acid, sandy soils, occasionally on more basic soils. In Britain only known in association with Scots pine (*Pinus sylvestris*).

Distribution and frequency: found throughout most of Europe and North America, and also reported from parts of Asia. However, the species is generally scarce and is considered endangered or extinct in some areas of Europe. In Britain it is rare, and found only in native Scottish pine woods.

Other remarks: the orange colours present throughout the fruitbody are characteristic and, together with the lack of clamp-connexions are diagnostic amongst British species. *Hydnellum caeruleum*, with orange stipe, may be similar to *H. aurantiacum*, but has bluish colours in the pileus and clamp-connexions present on the context hyphae.

Additional coloured illustrations of *H. aurantiacum* can be found in Breitenbach & Kränzlin (1986: pl. 257), Cetto (1979–84: 763, 1987a: 231), Marchand (1976: 341), Moser & Jülich (1985: V), Nilsson & Persson (1978: 81), and Ryman & Holmåsen (1984: 98).

Fig. 59. *Hydnellum aurantiacum* (Scotland, Inverness-shire, Abernethy Reserve, Nethybridge, G. Dickson 1012/90).

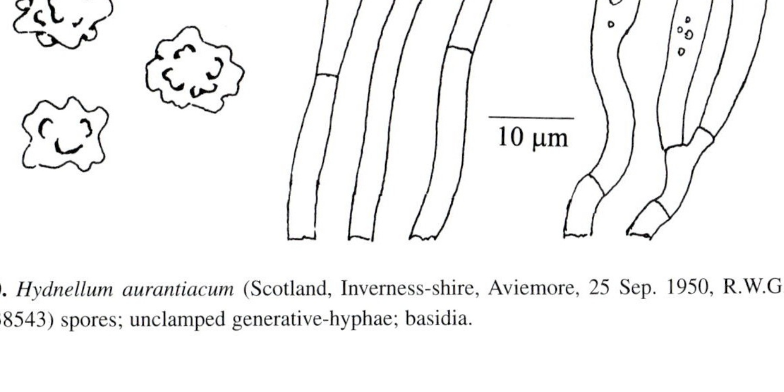

Fig. 60. *Hydnellum aurantiacum* (Scotland, Inverness-shire, Aviemore, 25 Sep. 1950, R.W.G. Dennis, K(M) 38543) spores; unclamped generative-hyphae; basidia.

4. CONCRESCENT CORKY SPINE FUNGUS

Hydnellum concrescens (Pers.) Banker in *Mem. Torrey Bot. Club* 12: 157 (1906).
Hydnum concrescens Pers., *Obs. myc.* 1: 74 (1796).
Hydnum concrescens Fr., *Epicr. Syst. Mycol.*: 509 (1838).
Hydnellum zonatum (Fr.) P. Karst. in *Medd. Soc. Fauna Fl. fenn.* 5: 41 (1838).
Calodon zonatus (Fr.) P. Karst. in *Bidr. Känn. Finl. Nat. Folk* 37: 108 (1882).
Phaeodon zonatus (Fr.) J. Schroet. in *Kryptog. Fl. Schles.* 3(1): 458 (1888).
Hydnellum velutinum var. *zonatum* (Fr.) Maas Geest. in *Fungus* 27: 64 (1957).
Hydnellum scrobiculatum var. *zonatum* (Fr.) K. Harrison, *Stip. Hydn. Nova Scotia*: 43 (1961).

Fruitbody usually concrescent, occasionally solitary. *Pileus* 2–8 cm across, soon depressed, irregularly lobed, reddish-brown to yellow-brown, sometimes with blackish patches, and with margin whitish at least when young, tomentose becoming smooth, with distinct radial grooves and ridges and with concentric zonation. *Stipe* usually central, variable in form, up to ca 5 cm long, 2–10 mm diam., cylindric or tapered, often with bulbous base, velutinous, concolorous with pileus, blackening when bruised. *Spines* decurrent, (1–)2–3(–4) mm long, slender, rusty-brown becoming pale buff. *Context* thin, reddish-brown, zoned, often darker in the stipe, in alkali solution becoming rapidly dark green, and exuding a green or blue-green stain; *odour* weak, of meal; *taste* slightly bitter. *Context hyphae* 4–6 µm diam., pale yellow-brown to hyaline, or with pale yellowish pigment, often branched, septate, clamp-connexions lacking; *hyphae in spines* narrow, (2–)3–3.5(–4) µm diam., often branched, flexuous, yellow-brown, remotely septate, lacking clamp-connexions. *Spores* 4.5–5.5 × 3–3.5 µm, yellow-brown, tuberculate. *Basidia* 29–37 × 5–6.5 µm, narrowly clavate, 4-sterigmate.

Habitat: woodlands, associated with both deciduous and coniferous trees.

Distribution and frequency: *Hydnellum concrescens* is widespread in the northern hemisphere, and found throughout Europe, commonly in some areas. It is also present in North America and much of Asia. In the British Isles, including Ireland, it is widely distributed, but is apparently most frequent in parts of Scotland and in southern England.

Other remarks: this species is closely related to *H. scrobiculatum*, having fruitbodies of a similar form and colour. It differs most notably in its more distinctly zoned and radially grooved cap and smaller spores. The distinct dark green stain which exudes from the flesh in alkali solution is also a distinguishing character.

For additional coloured illustrations see Breitenbach & Kränzlin (1986: pl. 261), Cetto (1979–84: 765, 1987a: 229, 1987b: 2037), Phillips (1981: 244), Moser & Jülich (1985: V), and Ryman & Holmåsen (1984: 99).

Fig. 61. *Hydnellum concrescens* (England, Hampshire, New Forest, East End, G. Dickson 327/85).

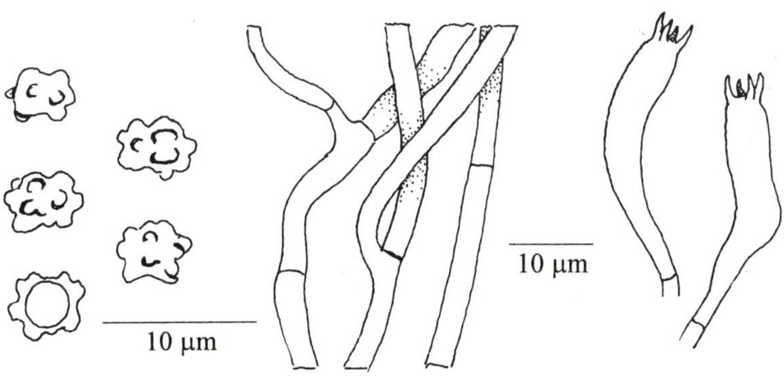

Fig. 62. *Hydnellum concrescens* (England, Hampshire, New Forest, 2 Nov. 1991, T. Laessøe, K(M) 18395) spores; unclamped generative-hyphae; basidia.

5. PITTED CORKY SPINE FUNGUS

Hydnellum scrobiculatum (Fr.) P. Karst. in *Medd. Soc. Fauna Flora fenn.* 5: 41 (1879).
Hydnum scrobiculatum Fr., *Obs. mycol.* 1: 143 (1815).
Hydnellum velutinum var. *scrobiculatum* (Fr.) Maas Geest. in *Fungus* 27: 63 (1957).
Calodon scrobiculatus (Fr.) P. Karst. in *Bidr. Känn. Finl. Nat. Folk* 37: 108 (1882).
Calodon zonatus var. *scrobiculatus* (Fr.) Quél., *Ench. Fung.*: 190 (1886).
Phaeodon scrobiculatus (Fr.) P. Henn. in *Nat. Pflanzenfam.* 1 (1**): 148 (1898).

Fruitbody solitary or caespitose to concrescent, stipitate. *Pileus* 3–5 cm across, rather thin-fleshed, depressed, finely velutinous becoming smooth, scrupose, with irregular coarse scales and excrescences especially at the centre, radially ridged and streaky and with weak concentric zones in shades of brown to grey-brown; sometimes becoming blackish where bruised. *Stipe* central or excentric, 1–3 cm long, 3–10 mm thick, cylindric or tapered, often with bulbous base, surface thickly velutinous and often binding debris, concolorous with pileus. *Spines* decurrent, 2–3(–4) mm long, slender, brown to purple-brown. *Context* thin, red-brown or purplish brown, often with blue-grey streaks and whitish spots. *Odour* of meal; *taste* mild or slightly spicy. *Context hyphae* 3–5(–6) µm diam., pale yellow-brown, thin-walled, straight to flexuous, sometimes branched, lacking clamp-connexions; *hyphae in spines* 2.5–3.5 µm diam., yellowish-brown, branched, remotely septate. *Spores* subglobose, pale yellowish-brown, tuberculate, 5–6.5 × 4–5.5 µm. *Basidia* 23–29 × 5–6.5 µm, narrowly clavate, 4-spored.

Habitat: coniferous and mixed woodland; able to grow with many trees, in Britain mainly associated with pine (*Pinus*) and *Fagaceae*.

Frequency & distribution: widely distributed in the northern hemisphere, and also present in Australasia. It occurs throughout much of Europe and North America, and parts of Asia, and is comparatively common in some areas. In the British Isles it is widely distributed but uncommon.

Other remarks: although somewhat variable, this species can usually be readily distinguished by the thin flesh and radially grooved pileus. *Hydnellum concrescens* is closely related and similar in many respects, but differs in having smaller spores, in the pileus usually having more distinct concentric zonation, and in rapidly exuding a dark greenish stain from the context in alkali solution. Further discussion on these two species is provided by Baird (1986b).

Additional coloured illustrations of *H. scrobiculatum* are to be found in Breitenbach & Kränzlin (1986: pl. 265) and Phillips (1981: 243).

Fig. 63. *Hydnellum scrobiculatum* (England, Surrey, Esher Common, 24 Aug. 1980, P.M. Kirk).

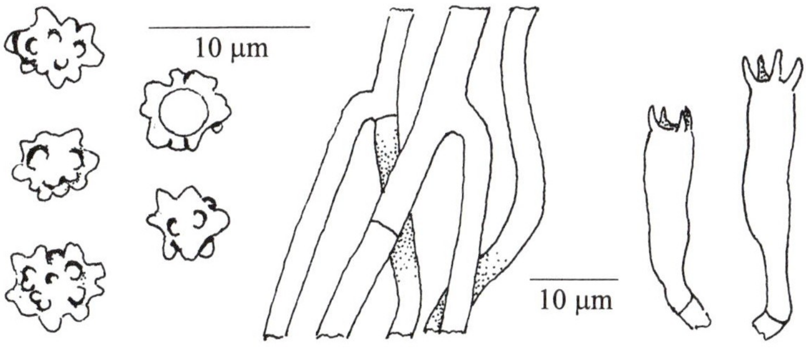

Fig. 64. *Hydnellum scrobiculatum* (England, Surrey, Windsor Great Park, 13 Sep. 1965, E.E. Green, K(M) 38649) spores; unclamped generative-hyphae; basidia.

6. REDDISH-BROWN CORKY SPINE FUNGUS

Hydnellum ferrugineum (Fr.: Fr) P. Karst. in *Medd. Soc. Fauna Flora fenn.* 5: 41 (1879).
Hydnum ferrugineum Fr., *Obs. mycol.* 1: 133 (1815); *Syst. Mycol.* 1: 403 (1821).
Calodon ferrugineus (Fr.: Fr.) P. Karst. in *Rev. Mycol.* 3(9): 20 (1881).
Phaeodon ferrugineus (Fr.: Fr.) J. Schroet. in *Kryptog. Fl. Schles.* 3(1): 459 (1888).
Phellodon ferrugineus (Fr.: Fr.) R. Nav. in *Natuurwet. Tijdschr., Antw.* 5: 68 (1923).

Fruitbody simple or usually concrescent, short stipitate. *Pileus* 3–10 cm across, turbinate, irregularly convex or pulvinate then plane to depressed, usually lacking radial markings, occasionally with concentric grooves, whitish or with pinkish tints, darker with age, often reddish-brown to dark brown, velutinous, becoming hispid or scrobiculate; developing fruitbodies commonly exuding red droplets in wet conditions. *Stipe* usually central, 2–5 cm high, 1–2 cm thick, cylindric or tapered, brown to reddish-brown, surface spongy-velutinous, binding debris. *Spines* decurrent, to c. 5 mm long, slender, 200–300 µm thick at base, whitish at first then darker, red-brown to purplish-brown with age, buff to pale brown in dried material. *Context* fawn-brown to pinkish-brown, darker below, with weak irregular darker zonation, and often with small whitish flecks, exuding a pale brownish stain in alkali solution; *taste* mild; *odour* farinaceous. *Context hyphae* 4–6(–7) µm diam., brownish, thin-walled, lacking clamp-connexions; *hyphae in spines* narrow, 3.5–4.5 µm diam., sometimes branched, thin-walled, septate, with long segments. *Spores* 5.5–7.5 × 4.5–5.5 µm, broadly ellipsoid to subglobose, pale yellowish, tuberculate. *Basidia* 25–30 × 6–7.5 µm, narrowly clavate, 4-spored.

Habitat: coniferous woodland, with pine (*Pinus*) or occasionally spruce (*Picea*), mostly on poor, sandy soil.

Frequency and distribution: widely distributed and present throughout most of Europe, but scarce; also known from North America and North Africa. Rare in Britain, known from the Scottish Highlands, parts of south and south-east England, the Welsh borders and southern Ireland.

Other remarks: closely related to *H. spongiosipes*; the two species are similar in many respects and have been much confused in the past. *Hydnellum spongiosipes* differs most notably in having darker flesh and occuring in deciduous woods. Blood-red droplets are commonly exuded in damp conditions from developing specimens of both species. Such droplets also occur in *H. peckii*, which is distinguished by the presence of clamp-connexions and its acrid taste.

For additional coloured illustrations see Breitenbach & Kränzlin (1986: pl. 262), Cetto (1987a: 225), Dähncke (1993: 1044), Marchand (1976: 344), and Ryman & Holmåsen (1984: 98).

Fig. 65. *Hydnellum ferrugineum* (Finland, Patkeljärri National Park, Aug. 1996, E.E. Emmett).

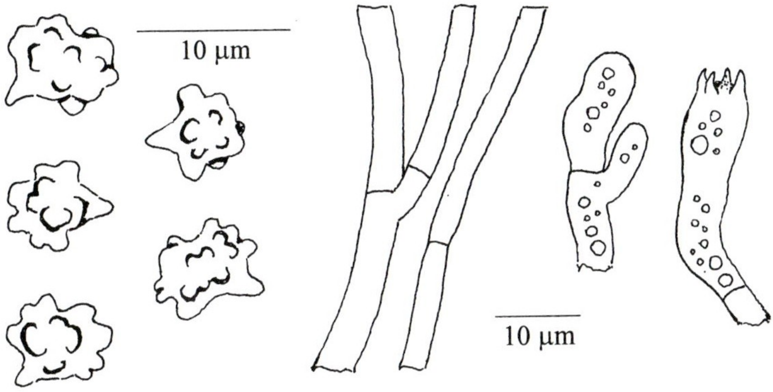

Fig. 66. *Hydnellum ferrugineum* (Wales, Gwynedd, Betws-y-Coed, 25 Sep. 1924, K(M) 38636) spores; unclamped generative-hyphae; basidia.

7. VELVETY CORKY SPINE FUNGUS

Hydnellum spongiosipes (Peck) Pouzar, in *Ceská Mykol.* 14: 130 (1960).
Hydnum spongiosipes Peck, in *Rep. N.Y. St. Mus. nat. hist.* 50: 111 (1897).
Hydnellum velutinum var. *spongiosipes* (Peck) Maas Geest., in *Fungus* 27: 62 (1957).

Fruitbody simple or clustered to concrescent, stipitate. *Pileus* 3–10 cm across, convex or irregularly pulvinate, becoming flattened or slightly depressed at the centre, lacking radial structures but oten with concentric grooves, tomentose or velutinous, becoming smooth, pale clay-brown at first, becoming darker, purple-brown or reddish-brown with age and where bruised, often exuding reddish droplets when fresh, margin often whitish when young. *Spines* decurrent, 2–5 mm long, slender, dark pinkish-buff to reddish-brown. *Stipe* rooting, 2–4(–6) × 1–2.5 cm, concolorous with pileus, bulbous, tapered below, surface thickly tomentose. *Context* rather dark, especially in stipe, brown to pinkish-brown or red-brown, zoned, often rather obscurely so, and often with small whitish spots; *odour* farinaceous; *taste* mild. *Context hyphae* 5–7 µm diam., straight to flexuous, sometimes branched, thin-walled, pale brown, clamp-connexions absent; *gloeoplerous hyphae* well developed, distinct, containing brown pigment; *hyphae in spines* narrow, mostly 3–4 µm diam., often branched, thin-walled, pale brown or yellow-brown, remotely septate. *Spores* subglobose, 6–7 × 5–5.5 µm, tuberculate, pale yellow-brown. *Basidia* 25–34 × 7–8 µm, clavate, (2–)4-spored.

Habitat: on sandy soil in deciduous woods, especially with oak (*Quercus*), beech (*Fagus*) and sweet chestnut (*Castanea*), rarely in mixed woodland but apparently never directly associated with conifers.

Frequency & distribution: present throughout most of Europe and also found in North America, but uncommon or rare and considered endangered in some areas. In Britain it is scarce, occurring mostly in southern England and East Anglia. It is also known from Ireland, but there appear to be no records from Scotland.

Other remarks: closely related to *H. ferrugineum*, and similar in many respects to that species. *Hydnellum ferrugineum* differs most notably in its paler, more distinctly zoned context and occurrence in coniferous woods. Further discussion of these two species is given by Baird (1986b).

Additional coloured illustrations of *H. spongiosipes* are to be found in Breitenbach & Kränzlin (1986: pl. 266), Cetto (1987b: 2036), Marchand (1976: 346), Moser & Jülich (1985: V), and Phillips (1981: 244).

Fig. 67. *Hydnellum spongiosipes* (England, Hampshire, New Forest, East End, G. Dickson 854).

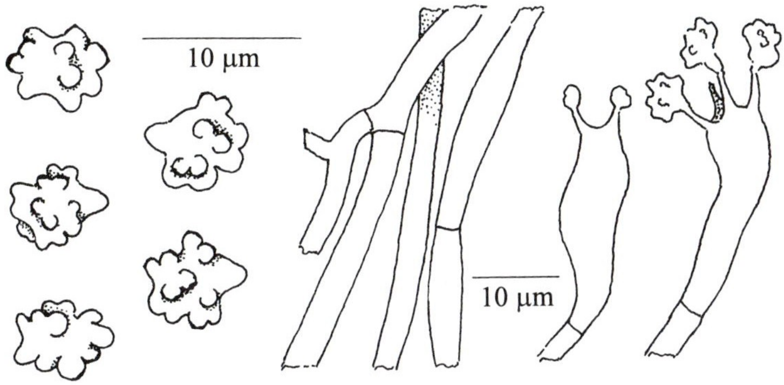

Fig. 68. *Hydnellum spongiosipes* (England, Berkshire, Swinley Park, 15 Sep. 1967, E.E. Green, K(M) 38909) spores; unclamped generative-hyphae; basidia.

Fig. 69. *Hydnellum aurantiacum* (Scotland, Inverness-shire, Coy, Lynwilg, 22 Sep. 1900, C. Rea), from an original painting by Carlton Rea.

2. SARCODON P.Karst.

in *Rev. mycol.* 3: 20 (1881).

Fruitbody annual, terrestrial, pileate, stipitate, simple or concrescent. *Pileus* fleshy, smooth to velutinous or scaly, brownish. *Stipe* central, rarely excentric, concolorous with pileus or paler, sometimes blue-green at base, tomentose. *Hymenium* covering spines on the underside of the pileus. *Spines* at first whitish, becoming brownish at maturity. *Context* soft or firm, brittle, not zoned, whitish or with brown to reddish or violaceous, or rarely green colours, sometimes blackish in stipe base. *Hyphal system* monomitic, of thin-walled, branched, septate, often inflating generative hyphae, clamp-connexions present or not. *Basidia* clavate, 4-spored; basal clamp-connexion present or not. *Spore deposit* brown. *Spores* subglobose to ellipsoid, yellow-brown or brownish, irregularly tuberculate. *Cystidia* lacking. *Type species*: *Hydnum imbricatum* L.: Fr.

The genus is widely distributed in the northern hemisphere, occurring throughout much of Europe, Asia and North America; it is also present in Australasia. Species are ectomycorrhizal, occurring with both angiosperms and gymnosperms. Sixteen species are recognized in Europe by Maas Geesteranus (1975), many of them now rare or endangered and several regarded as extinct in some parts of their range. In the British Isles, five species are recognized in the present study, although the presence here of *S. fuligineoviolaceus* requires confirmation. Amongst these, *S. imbricatus* is perhaps the best known and least rare but nevertheless regarded as vulnerable by Ing (1992). The others are very rare, and apparently restricted in distribution, and are regarded either as endangered or, in the case of *S. regalis*, as extinct in Britain (Ing 1992). *Sarcodon leucopus* (Pers.) Maas Geest. & Nannf. has also been recorded as British, but records appear to be based on errors of identification and there is as yet no good evidence of its presence in the British Isles.

Identification of species of *Sarcodon* requires specimens in fresh condition so that the colour of the flesh, both in pileus and stipe, can be ascertained. This, together with the presence or absence of clamp-connexions on context hyphae, are the most helpful characters in delimitation of species.

Key to Species

1. Clamp-connexions present on context hyphae; pileus with conspicuous, coarse concentric scales .... **1. Scaly Tooth** (*S. imbricatus*)
1. Clamp-connexions absent on context hyphae; pileus scaly or not **2**

  2. Context of pileus dark blue-violet when cut, reddish in stipe .... **2. Burnt Spiny Cap-Fungus** (*S. fuligineoviolaceus*)
  2. Context of pileus whitish or at most with reddish-violet tinge when cut; stipe flesh violaceous or blue-green to grey-green at base **3**

3. Stipe with violaceous tints; pileus with scales adpressed or with raised tips .... **3. Regal Spiny Cap-Fungus** (*S. regalis*)
3. Stipe lacking violaceous tints; pileus with scales adpressed or erect at centre **4**

  4. Pileus with scales erect at centre .... **5. Bitter Spiny Cap-Fungus** (*S. scabrosus*)
  4. Pileus smooth or scales adpressed **6. Green-footed Spiny Cap-Fungus** (*S. glaucopus*)

1. SCALY TOOTH

Sarcodon imbricatus (L.: Fr.) P. Karst. in *Rev. mycol.* 3: 20 (1881).
Hydnum imbricatum L.: Fr., *Syst. mycol.* 1: 398 (1821).
Phaeodon imbricatus (L.: Fr.) Schroet., *Kryptog. Fl. Schles.* 3: 460 (1888).

Fruitbody simple or sometimes connate, solitary or clustered. *Pileus* to 20 cm across, yellow-brown then darker, circular or irregular in outline, sometimes lobed, convex, centre usually depressed, eventually infundibuliform, prominently squamulose; squamules concentric, imbricate, coarse, erect at the centre, towards the margin more fibrillose. *Stipe* cylindric or somewhat tapered towards the base, 2–6 cm long, 1–3(–5) cm diam., whitish at first, soon brownish, surface finely fibrillose or squamulose becoming smooth. *Spines* decurrent, 5–12 mm long, slender, subulate, crowded, whitish or pale at first, soon brown. *Context* to ca 3 cm thick, whitish or with reddish or vinaceous flush, darker in stipe base, lacking green tints; *odour* weak, slightly spicy; *taste* mild or slightly bitter. *Context hyphae* 8–20(–23) µm diam., hyaline, thin-walled, septate, with clamp-connexions present; *hyphae in spines* narrower, 3.5–9 µm diam., hyaline or faintly pigmented, septate with clamp-connexions, with long segments, often branched. *Spores* 6–8 × 5–6.5 µm (incl. tubercles), pale brownish, tuberculate; tubercles prominent, apically flat or indented. *Basidia* 25–30 × 5.5–7 µm, narrowly clavate, with basal clamp-connexion.

Habitat: coniferous woods, especially with pine (*Pinus*), on acid, sandy soil.

Distribution and frequency: the commonest species of the genus, occurring throughout Europe and temperate North America, and also present in eastern Asia. In Europe it is associated with conifers, including spruce (*Picea*) and pine (*Pinus*), but it may be noted that North American collections are frequently from hardwood forests (Coker & Beers 1951; Baird 1986b). Baird (1986b) notes uncertainty as to whether these are conspecific with European collections although they are morphologically indistinguishable. The species is widespread but rare and considered as vulnerable in Britain It is mostly eastern in distribution, occurring mainly in central and southern Scotland and in south-east England and East Anglia.

Other remarks: the species may be readily distinguished by the coarsely scaly pileus and presence of clamp-connexions. The rare *Sarcodon scabrosus* also has erect scales and is similar in appearance but occurs in deciduous woodland and differs otherwise in the lack of clamp-connexions and the presence of blue-green tints in the stipe base. *Sarcodon glaucopus*, also with conifers, has smooth or cracking pileus with adpressed scales and again lacks clamp-connexions.

Additional coloured illustrations of *S. imbricatus* are to be found in Breitenbach & Kränzlin (1986: pl. 275), Cetto (1979-84: 326, 1987a: 243), Dähncke (1993: 1045), Marchand (1976: 348), Moser & Jülich (1985: V), Nilsson & Persson (1978: 77), Phillips (1981: 242), and Ryman & Holmåsen (1984: 103).

Fig. 70. *Sarcodon imbricatus* (Scotland, Inverness-shire, Coylumbridge, Lairig Ghru, September, Dickson 799/85).

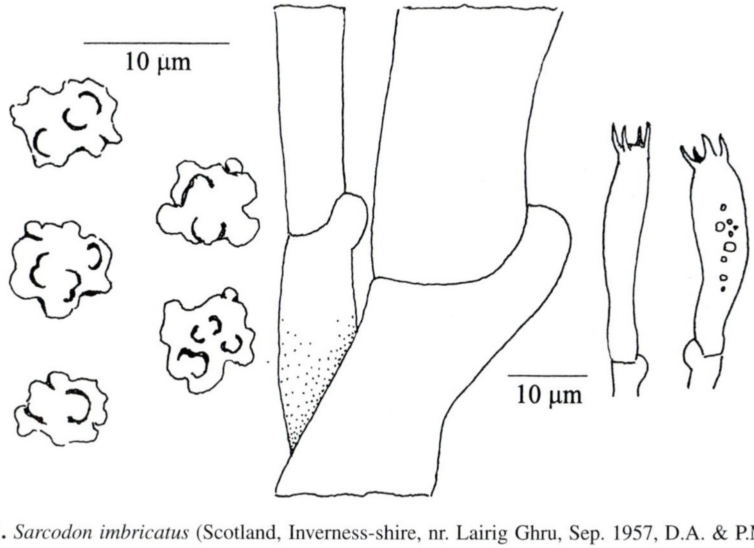

Fig. 71. *Sarcodon imbricatus* (Scotland, Inverness-shire, nr. Lairig Ghru, Sep. 1957, D.A. & P.M. Reid, K(M) 40689) spores; swollen context hyphae showing clamp-connexions; basidia.

2. BURNT SPINY CAP-FUNGUS

Sarcodon fuligineoviolaceus (Kalchbr. in Fr.) Pat., *Essai tax. Hym.*: 118 (1900).
Hydnum fuligineoviolaceum Kalchbr. in Fr., *Hym. europ.*: 602 (1874).
Sarcodon talpa Maas Geest. in *Persoonia* 5: 9 (1967).

Fruitbody usually simple, rarely concrescent. *Pileus* 3–13 cm across, yellow-brown to reddish-brown or with olive tints, finally with bluish or violaceous tinge, particularly at the centre, darkening with age, convex to flattened or depressed, at first tomentose, becoming fibrillose to scaly, without concentric markings; *scales* adpressed or with raised tips. *Stipe* 3–6 × 1–3 cm, cylindric or slightly tapered downwards with somewhat rooting base, tomentose to fibrillose, sometimes becoming rather scaly, purplish-brown or concolorous with pileus, sometimes with greenish tints at base, and with a whitish basal tomentum. *Spines* decurrent, to ca 4 mm long, slender, subulate, pinkish-brown or purple-brown, often with white tips. *Context* pinkish at first, then dark blue-grey to violaceous in pileus, becoming violet-grey when dried, reddish to vinaceous-red in stipe, grey-green in stipe base; staining blue-green in alkali solution; *taste* acrid. *Context hyphae* inflating, 6–15(–23) µm diam., often branched, thin-walled, lacking clamp-connexions; *hyphae in spines* thin-walled, sometimes branched, long-celled, 2–6 mm diam., not staining green in alkali solution. *Basidia* 35–53 × 6–8 µm, narrowly clavate, 4-sterigmate. *Spores* 5–6.5 × 4–5 µm, tuberculate, irregular in outline, pale brown.

Habitat: coniferous woods, mainly with pine (*Pinus*).

Distribution and frequency: recorded from various parts of northern Europe and North America, and also present in eastern Asia, though apparently uncommon or rare throughout its range. The British status of this species is unclear, and, although included and given as endangered by Ing (1992), its presence here requires confirmation. The species is recorded as British by Jülich (1984), but no records have been traced during the present study, and no British material seen.

Other remarks: the above description is partly based on an examination of the Dutch specimen cited and partly adapted from Maas Geesteranus (1971, 1975); a further description, based on Russian material, is given by Nikolaeva (1977). The blue-violet colour in the pileus and the absence of clamp-connexions should serve to distinguish this species.

Additional coloured illustrations of *S. fuligineoviolaceus* appear in Cetto (1987: 2038) and in Maas Geesteranus (1975: pl. 38, 39a).

Fig. 72. *Sarcodon fuligineoviolaceus* (Norway, Elvestad, 24 Aug. 1995, P. Marstad).

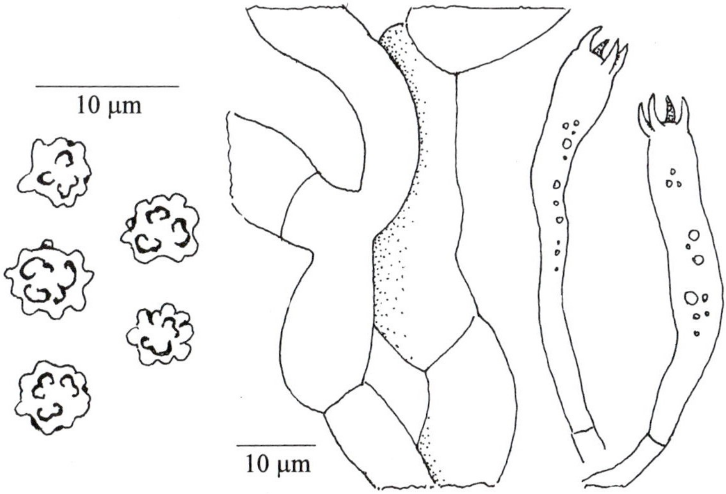

Fig. 73. *Sarcodon fuligineoviolaceus* (Netherlands, Gelderland, Wageningen-Hoog, 1 Oct. 1963, R.A. Maas Geesteranus 13966, K(M) 40701) spores; inflated hyphae in context; basidia.

3. REGAL SPINY CAP-FUNGUS

Sarcodon regalis Maas Geest., *Die Terrestrischen Stachelpilze Europas*: 106 (1975).

Fruitbody simple or concrescent. *Pileus* to ca 10 cm across, convex to flattened, sometimes lobed, scaly, fibrillose towards the margin, yellowish-brown sometimes with purplish tint. *Scales* at centre broad, adpressed or with raised tips, towards the margin narrower and fibrillose, grey-brown to dark brown. *Stipe* 3–6 × 1.5–2 cm, central or excentric, simple or occasionally branched, cylindric or usually tapered downwards, rooting, surface tomentose or fibrillose, pale orange-brown, with violaceous tints below, base white tomentose sometimes with grey-green marks. *Spines* decurrent, slender, subulate, 1.5–3.5 mm long, pale becoming brown to purplish brown. *Context* up to ca 10 mm thick in pileus, whitish, vinaceous to violaceous in patches, often pale reddish-brown in stipe, grey-green in stipe base; *odour* pleasant, soon fruity; *taste* farinaceous becoming bitter. *Context hyphae* 10–21 µm diam., hyaline, thin-walled, branched, lacking clamp-connexions; *hyphae in spines* narrow, 3–5 µm diam., septate, with long segments, sometimes branched, hyaline to pale yellow-brown. *Basidia* 33–38 × 5–7 µm, narrowly clavate. *Spores* 5–6.5 × 4–5 µm (incl. tubercles), subglobose, yellow-brown, tuberculate.

Habitat: typically associated with oak (*Quercus*) and sweet chestnut (*Castanea*); reported with conifers by Cetto (1987b).

Distribution and frequency: apparently very rare, and now regarded as extinct in Britain (Ing, 1992). It is known here only from the type material from Berkshire, Windsor Great Park, collected in October 1968 and in September 1969. The species has subsequently been recorded from Germany, and more recently from France (Jülich, 1984) and Italy (Cetto, 1987b).

Other remarks: this is the only Tooth-Fungus which has been described from the British Isles. The earlier of the two British collections cited in the original description was selected as lectotype and redescribed by Baird (1986a). The species is distinguished by the lack of clamp-connexions, the adpressed scales on the pileus and the strong violaceous tints in the stipe.

Additional coloured illustrations of this species appear in Cetto (1987b: 2040), and in Maas Geesteranus (1975: pl. 32 b–c, 33a).

Fig. 74. *Sarcodon regalis* (England, Berkshire, Windsor Great Park, Swinley Park, 3 Oct. 1968, R.A. Maas Geesteranus 15291); after R.A. Maas Geesteranus, 1976: pl. 32b.

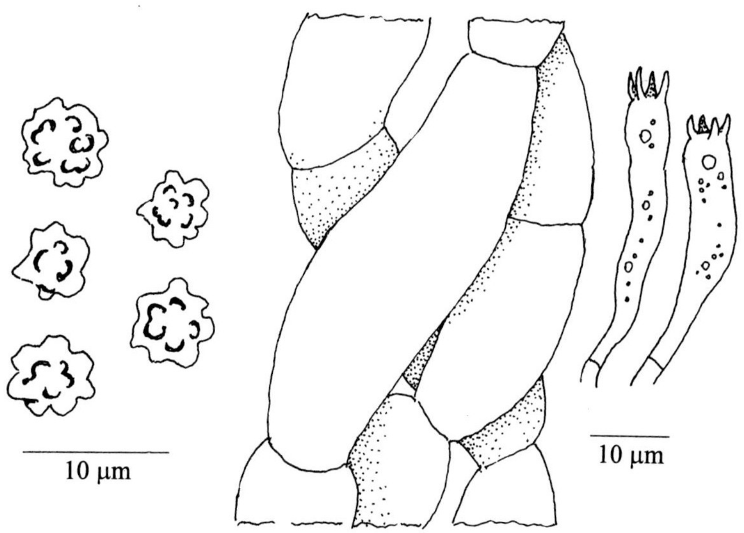

Fig. 75. *Sarcodon regalis* (England, Berkshire, Windsor Great Park, Swinley Park, 3 Oct. 1968, R.W.G. Dennis et al., K(M) 40694, holotype) spores; swollen context hyphae; basidia.

4. BITTER SPINY CAP-FUNGUS

Sarcodon scabrosus (Fr.) P. Karst. in *Rev. mycol.* 3: 20 (1881).
Hydnum scabrosum Fr., *Anteckn. Sverige väx. ätl. Svamp.*: 62 (1836).

Fruitbody simple or caespitose to concrescent. *Pileus* 5–15 cm across, plano-convex, becoming plane or depressed, brown, sometimes with pinkish tinge, often blackish with age, extreme margin whitish when young, surface soon splitting to form scales. *Scales* coarse, erect at centre, adpressed towards the margin, brown to reddish-brown, contrasting with paler background. *Stipe* 3–10 × 1–3 cm, cylindric or tapered below, fibrillose or finely scaly, pinkish-brown to dark brown, blue-green to greenish-black towards the base. *Spines* decurrent, 4–6(–10) mm long, pale greyish-pink, becoming purple-brown, tips whitish. *Context* whitish or with pinkish tinge, blue-green or grey-green at the stipe base; *odour* mealy; *taste* bitter to somewhat acrid. *Context hyphae* (9–)11–20(–23) μm diam., hyaline, thin-walled, lacking clamp-connexions; *hyphae in spines* narrow, 4–9 μm diam., thin-walled, yellow-brown, septate, with long segments. *Basidia* 40–60 × 6.5–8.5 μm, narrowly clavate, 4-spored. *Spores* 6–7.5 × 5–6 μm (incl. tubercles), broadly ellipsoid to subglobose, yellow-brown to brown, tuberculate.

Habitat: in coniferous and deciduous woods, usually on acid, sandy soils. In Britain recorded in the south mainly with *Fagaceae*, especially oak (*Quercus)* and chestnut (C*astanea)*, and in Scotland with Scots pine (*Pinus sylvestris*).

Distribution & frequency: recorded throughout much of Europe, but generally rare; in some areas it is regarded as threatened or vulnerable. The species is also widespread in temperate North America where, in contrast, it is regarded as common in places. It is also present in India. In Britain it is rare, and regarded as endangered by Ing (1992). It is recorded from the Scottish Highlands and from parts of south and south-east England.

Other remarks: fruitbodies are similar in appearance to those of *S. imbricatus* but differ especially in the blue-green or greenish-black colour at the base of the stipe and the lack of clamp-connexions. *Sarcodon scabrosus* is closely related to *S. glaucopus*, which also has grey-green tints in the stipe base, but the latter differs most notably in the adpressed scales of the pileus.

Additional coloured illustrations of *S. scabrosus* appear in Breitenbach & Kränzlin (1986: pl. 279), Moser & Jülich (1985: V), and Phillips (1981: 243).

Fig. 76. *Sarcodon scabrosus* (England, Berkshire, Windsor Great Park, D.A. Reid. Kew photograph 2486).

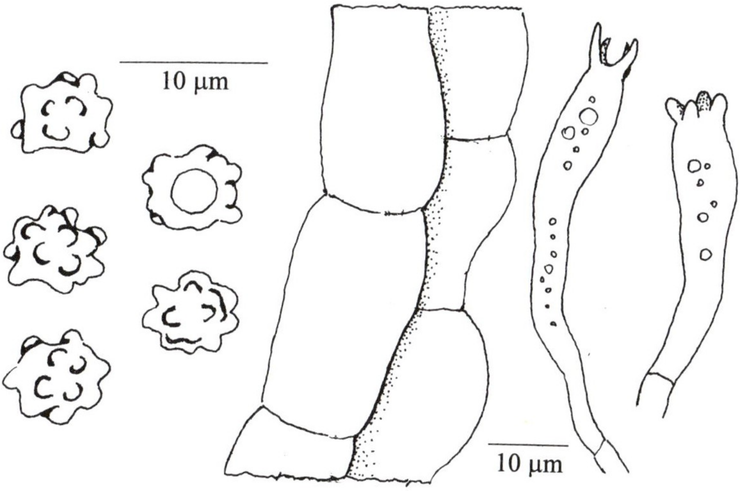

Fig. 77. *Sarcodon scabrosus* (England, Kent, Pembury, Tudeley Woods, 24 Sep. 1994, N. Fletcher, K(M) 30165) spores; swollen context hyphae; basidia.

5. GREEN-FOOTED SPINY CAP-FUNGUS

Sarcodon glaucopus Maas Geest. & Nannf. in *Svensk Bot. Tidskr.* 63: 407 (1969).

Fruitbody simple or fused at the margin, solitary or gregarious. *Pileus* 7–12 cm across, convex or expanded-convex, becoming plane, sometimes incurved and wavy-lobed at the margin, clay-buff or fawn, with paler buff to pale clay-pink margin, centre sometimes umber or tinged violaceous when fresh, surface matt, smooth or becoming radially finely to coarsely cracked or areolate especially in the centre, outer part with small adpressed scales then almost smooth. *Stipe* 20–50 × 10–25mm, usually somewhat tapered downwards but not rooting, apex with decurrent small spines, milky-coffee or fawn below spines, grey-olivaceous or greenish-grey at base, sometimes pale cinnmon or tinged tawny in the middle part, base sometimes tinged leaden grey, extreme base fuscous-black; finally blackish punctate below the spines. *Spines* slender, 2–3 mm long, whitish then vinaceous-buff or clay pinkish to fawn, tips often remaining whitish. *Context* in pileus and stipe apex pale buff, straw or cream, towards the base tinged olivaceous buff or greenish-grey, extreme base fuscous-black; *odour* strong mealy when cut; *taste* mild to slightly acrid. *Context hyphae* 5–22 µm diam., hyaline, thin-walled, simple septate, lacking clamp-connexions; *hyphae in spines* narrow, 3–5 µm diam., simple-septate, with long segments, sometimes branched. *Spores* (5.5–)6–7.5 × 4–5.5 µm (incl. tubercles), ellipsoid to subglobose, yellow-brown, tuberculate. *Basidia* 26–40 × 6.5–7 µm, narrowly clavate, 4-spored.

Habitat: in coniferous woods. In Britain known only from native Scots pine (*Pinus sylvestris*) forest, growing especially at path edges.

Distribution and frequency: although recorded from much of Europe, it is apparently rare throughout and is regarded as extinct in Poland and as threatened with extinction in Germany. It is not recorded from North America. In Britain it is very rare, being hitherto known only from Abernethy Forest in Scotland.

Other remarks: *Sarcodon glaucopus* is characterized by the lack of clamp-connexions, and the lack of violaceous tints in the flesh of the pileus. The surface of the pileus may be smooth but commonly cracks to form adpressed scales. The absence of erect scales on the pileus and the habitat distinguish it from the closely related *S. scabrosus*.

This species was first collected in Britain in 1990 by P.D.Orton. It was recollected in 1991 and 1992, and is now known from two sites ca 1 km apart in native Caledonian pine woods.

An additional coloured illustration of *S. glaucopus* appears in Cetto (1987a: 249).

Fig. 78. *Sarcodon glaucopus* (Scotland, Inverness-shire, Abernethy Reserve, Nethybridge, G. Dickson 989/90).

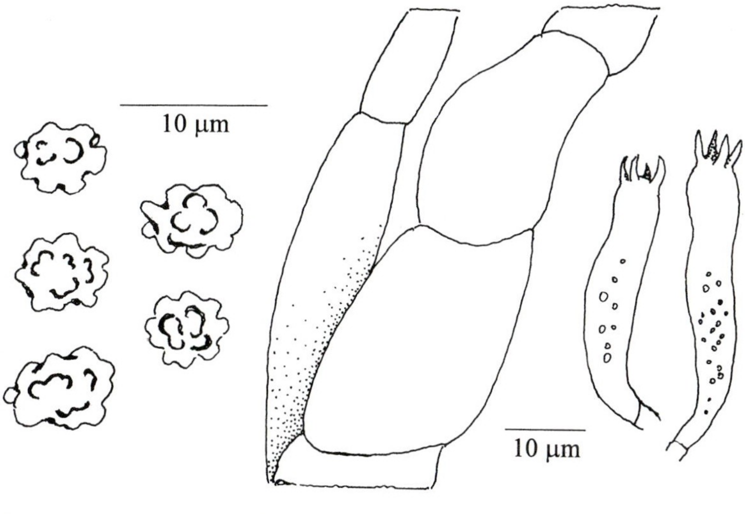

Fig. 79. *Sarcodon glaucopus* (Scotland, Inverness-shire, Abernethy, 5 Sep. 1991, P.D. Orton, K(M) 18284) spores; swollen context hyphae; basidia.

GLOSSARY

adnate (of gills, tubes): broadly attached to stipe apex.
adnexed (of gills, tubes): narrowly attached to stipe apex.
adpressed (of scales, etc): lying flat against the surface.
agaric: a general term for mushrooms and toadstools i.e. order *Agaricales*.
agaricoid: referring to a fruitbody which resembles a mushroom in having lamellae.
aliphatic: denotes organic compounds with open rings; fatty
amyloid (of spores and hyphae): turning grey to violaceous in Melzer's reagent (iodine reaction).
anastomosing (of lamellae, hymenial ridges, etc.): cross-connecting to form a network.
aphyllophoroid: resembling a group of basidiomycetes, formerly placed in the *Aphyllophorales*, which lack lamellae, eg. bracket, hydnoid, club and coral fungi.
areolate (of surfaces): divided by superficial cracking into small angular compartments.

ballistosporic: producing actively projected spores.
basidiomycete: a fungus producing basidia.
basidiospore: reproductive 'cell', resulting from meiotic division of the nucleus, produced by a basidium.
basidium (-ia): the reproductive unit, a single 'cell' in which nuclear fusion and meiosis occur, producing spores, typically four but may be 1–8 , at the apex.
bicyclic: forming two circles.
binding hyphae (= ligative hyphae): thick-walled, non-septate, branching hyphae, originating from generative hyphae.

caespitose (of fruitbodies): in groups or tufts.
carotene: orange-coloured pigment.
chiastic (of basidium): meiotic spindle transverse to longitudinal axis.
clamp-connexion (of hyphae): a characteristic structure at the septum, with a curved outgrowth growing out from one segment and by-passing the septum before reuniting with the next segment, allowing transfer of nucleus.
clavarioid: resembling a group of fungi having club- or coral-shaped fruitbodies.
clavate: club-shaped.
concrescent (of fruitbodies): fused together.
connate (of stipes): fused together at the base.
context: flesh of a fruitbody, which does not contain spore-producing units.
convex (of pileus): broadly rounded upwards in outline.
coriaceous: tough and leather-like.
cutis: (of pileus or stipe) surface layer.
cystidiole: small hymenial cystidium, often no more than a sterile basidium.
cystidium (-ia): sterile, often distinctively shaped element found in the hymenium or pileal and stipe surfaces.

decurrent (of gills, tubes): broadly attached and extending down the stipe.
dextrinoid (of spores and hyphae): turning reddish brown to crimson in Melzer's reagent (iodine reaction).
dimitic (of hyphal systems): having either skeletal hyphae or ligative hyphae, in addition to generative-hyphae.
duplex (of context): two-layered.

echinulate (of spores): having very small spines.
ectomycorrhiza (-ae): mycorrhiza in which hyphae externally wrap around root filaments. Ectomycorrhizae are typically formed by larger basidiomycetes, including many agarics, Chanterelles, and Tooth-Fungi.
epigeous (of development): growing or developing at ground level.
eu-hymenium: hymenium forming a palisade; arising from a clearly defined subhymenial layer.
expanded (of pileus): opening fully at maturity.

farinaceous: flour-like.
fascicle (of hyphae): a cluster.
fertile surface: spore-producing surface.
fibril: a small fibre.
fibrillose: bearing fibrils.
flabelliform: (of pileus) fan-shaped.
floccose (of pileus, stipe): bearing matted, woolly hairs.
fruitbody: a compound structure containing the spore-producing units of a fungus.
free (of gills, tubes): not reaching the stipe apex.
fuliginous: sooty colour.
furcate: (of lamellae or hymenophore) forked.
fuscous: brown tinged with grey or black.

generative-hyphae (of hyphal systems): basic hyphae of unlimited growth, branching, septate, thin- or thick-walled, with or without clamp-connexions.
gibbous: (of pileus) with an excentric umbo.
glabrous (of a surface): smooth, neither hairy nor scaly.
globose: ball-shaped.
gloeiferous (of hyphae): gloeoplerous; having oily contents.
gloeocystidium (-ia): cystidium with oily contents.
gloeoplerous (of hyphae): having oily (oleaginous) contents
gregarious (of fruitbodies): growing as a group but not connected to each other.
guttulate (of spores and basidia): containing one or more oil globules.
guttule: an oil globule.
gymnocarpic (of fruitbody development): referring to fruitbodies in which the spore-producing region is exposed at all stages of development.

hispid: covered with erect, stiff hairs.
holothecial (of fruitbody): hymenium covering all surfaces; no differentiation between fertile and sterile surfaces.
homobasidiomycete: a basidiomycete with undivided basidia.
homoiomerous (of context): composed exclusively of hyphae.
humicolous: growing in or on soil
hygrophanous (of pileal surface and context): watery in appearance, with water disappearing rapidly on drying.
hymenium: the palisadic layer of basidia, covering the hymenophore.
hymenophoral trama: that part of the context which supports the subhymenial layer and hymenium.
hymenophore: spore-bearing surface layer of the fruitbody.
hydnaceous: with a spiny or toothed hymenophore.

hydnoid (of hymenophore): composed of conspicuous teeth or spines.
hypha (-ae): individual filament which forms the basic structure of both the mycelium and fruitbody.

imbricate (of fruitbodies): overlapping each other.
inamyloid: negative reaction to Melzer's reagent, see amyloid.
infundibuliform: funnel-shaped.

lamellae: lower, vertical and radial plates or 'gills' of agaricoid fruitbodies.
lamellate (of hymenophore): composed of radial lamellae ('gills'), resembling a mushroom.
ligative hyphae (= binding hyphae): thick-walled, non-septate, branching hyphae, originating from generative-hyphae.
lignicolous: growing in or on wood.

meiosis: reduction division of the chromosomes.
meiotic spindle: a framework of fibres formed between the centrioles during nuclear division at meiosis.
membranous (of context): thin, not fleshy.
merulioid (of fertile surface): wrinkled and folded, *Merulius*-like.
mesopodal: with the stipe placed centrally.
micron (μm): one-thousandth of one millimetre.
monomitic (of hyphal systems): only one hyphal type present i.e. generative-hyphae.
mycelium (-ia): mass of hyphae forming the vegetative thallus, normally permeating the substrate e.g. soil, rotten wood, leaf litter.
mycorrhiza (-ae): the hyphae of a fungus forming a mutualistic relationship with the roots of a plant or tree; the relationship may be facultative or obligatory.

oleiferous (of hyphae): with oily contents, also termed 'gloeoplerous'.

papillate: with a small, nipple-shaped elevation.
perforate (of pileus): with a central opening, continuous with the hollow stipe.
peridiopellis: surface layer of the pileus.
pervious (of pileus): with a central opening.
pileate: bearing or forming a pileus.
pileipellis: the surface layer of the pileus.
pileus: the sterile cap of a fungal fruitbody.
pilothecial (of fruitbody): consisting of a pileus and a stipe.
pleuropodal: with the stipe placed laterally.
plicate: folded like a fan.
poroid (of hymenophore): formed of vertical tubes, each opening by a terminal pore.
primordium: initial stage in fruitbody formation.
pseudostipe: stem-like base which is not structurally distinct from the upper part of fruitbody.
pulvinate: cushion-like.
punctate: covered with fine raised dots.

reniform (of spores): kidney-shaped.
resupinate (of fruitbody): attached by the back surface to the substratum.
rhizomorphs: cord-like strands of mycelium, resembling plant roots.

rugose: wrinkled.
rugulose: finely wrinkled.

saprophytic: growing on dead organic matter.
scrobiculate (of a surface): marked with pits or depressions.
scrupose (of as surface): covered with rough, jagged points.
septum (-a): the cross-walls separating hyphal segments.
skeletal hyphae (of hyphal systems): thick-walled, non-septate, non-branching hyphae of limited growth, originating from generative-hyphae.
spathulate (of pileus): spatula-shaped, with a tapering base.
spore deposit: the pattern made by large numbers of spores which have been actively released from the fertile surfaces (gills, spines, tubes, etc).
squamose: forming scales.
squamulose: forming small scales
stereoid (of fruitbody): resembling a *Stereum* fruitbody.
sterigma (-ata): slender apical extension of the basidium, which bears the basidiospore.
stichic (of basidium): meiotic spindle parallel to longitudinal axis.
stipe: structurally distinct stem of the fruitbody.
stipitate (of fruitbody): possessing a stipe ('stem').
stirps: a group of species, below the rank of Section.
striate (of pileus): with minute, radiating furrows or lines, especially near the margin.
subhymenial layer: a defined region which produces the basidia to form the hymenium.
subiculum: a thin layer of mycelium on the substratum, from which the resupinate fruitbody arises.
subulate: awl-shaped; tapering from base to apex.

terricolous: living on the soil.
thickening hymenium: new basidia continuously added and growing between and beyond older basidia; see eu-hymenium.
tomentose: having a covering of dense, fine hairs.
torulose: tubular, but with many small bulges and constrictions.
trama: the sterile layer within the hymenophore.
trichodermium (of pileal surface): formed by a layer of short, vertical hyphae.
tuberculate: with wart-like projections.
turbinate: top-shaped, like an inverted cone.

umbilicate (of pileus): having a central, navel-like depression.
umbo (of pileus): a distinct, raised area at the centre of the pileus.
umbonate (of pileus): having an umbo.
urniform (of basidium): short and swollen towards the base.

veil layers: protective tissues covering the young fruitbody.
velutinous: velvety, densely covered with fine short hairs.
ventricose: swollen or enlarged in the middle.
verrucose: warted.
verruculose: finely warted.

REFERENCES

Arnolds, E. (1989). Former and present distribution of stipitate hydnaceous fungi (basidiomycetes) in the Netherlands. *Nova Hedwigia* 48: 107–142.

Badham, C.D. (1847). *A treatise on the esculent funguses of England*. Pp. 138. London: Reeve, Brothers.

Baird, R.E. (1986a). Type studies of North American and other related taxa of stipitate hydnums: Genera *Bankera, Hydnellum, Phellodon, Sarcodon. Bibl. Myc.* 103: 89 pp.

Baird, R.E. (1986b). Study of the stipitate hydnums from the Southern Appalachian Mountains Genera: *Bankera, Hydnellum, Phellodon, Sarcodon. Bibl. Myc.* 104: 156 pp.

Berkeley, M.J. (1860). Mycology V & XIII. *Gard. Chron. & Agric. Gaz.* 1860: 121 & 312.

Bolton, J. (1788). *An History of Fungusses growing about Halifax* 1: 73 pl. Halifax.

Bramley, W.G. (1985). *A fungus flora of Yorkshire 1985*. Pp. 277. Leeds: Yorkshire Naturalists' Union

Breitenbach, J. & Kränzlin, K. (1986). *Fungi of Switzerland* 2. *Non-gilled fungi*. Pp. 412. Lucerne: Verlag Mykologia.

Bresadola, J. (1929). *Iconographia mycologica* 10: tab. 451–500. Mediolani.

Cetto, B. (1979–84). *Der grosse Pilzführer*. Band 1-4. BLV Verlag.

Cetto, B. (1987a). *Enzyklopädie der Pilze* 1: Pp. 662. München: BLV Verlag.

Cetto, B. (1987b). *I funghi dal vero*. Vol. 5. Trento: Saturnia.

Clusius, C. [Charles de l'Ecluse] (1601). *Fungorum in Panonia observatorum brevis historia.*

Coker, W.M. & Beers, A.H. (1951). *The stipitate hydnums of the eastern United States.* University of North Carolina.

Comitato di Studio del Gruppo di Vicenza (1989). Le *Cantharellaceae* del Vicentino. *Riv. Mycologia* 32: 212–225.

Cooke, M.C. (1862). *A plain & easy account of British fungi*. Pp. 148. London: Robert Hardwicke.

Cooke, M.C. (1888). *Illustrations of British fungi (hymenomycetes)* 7: pl. 922–1101. London: Williams & Norgate.

Cooke, M.C. (1902). *Edible & poisonous mushrooms*. Pp. 127. London: SPCK.

Corner, E.J.H. (1957). *Craterellus* Pers., *Cantharellus* Fr. and *Pseudocraterellus* gen. nov. *Sydowia, Beih.* 1: 266–276, figs. 1–7.

Corner, E.J.H. (1966). A monograph of the cantharelloid fungi. *Ann. Bot., Memoir* 2: 255 pp, 5 pl., 134 text-figs.

Corner, E.J.H. (1970). Supplement to 'a monograph of *Clavaria* and allied genera'. *Beih. Nova Hedw.* 33: 299 pp.

Courtecuisse, R. (1993). Macromycètes intéressants, rares or nouveaux VII. Taxas nouveaux pour la France et autres récoltes remarquables. *Docum. Mycol.* 23 (fasc. 91): 1–12, pl. 3.

Dähncke, R.M. (1993). *1200 Pilze in Farbfotos*. Pp. 1179. Stuttgart: AT Verlag.

Damblon, J., Demoulin, V. & Schumacher, R. (1974). *Cantharellus melanoxeros* Desm. *Bull. Soc. Mycol. Fr.* 90: 5–8.

Damblon, J., Demoulin, V. & Schumacher, R. (1975). *Cantharellus melanoxeros* Desm. *Bull. Soc. Mycol. Fr.* 91: Atl. pl. 199.

Dannell, E. (1994). *Cantharellus cibarius*: mycorrhiza formation and ecology. *Comp. Summaries of Uppsala Dissert. from Fac. Sci. Tech.* 35: 75 pp.

Donk, M.A. (1933). Revision der Niederländischen *Homobasidiomycetae-Aphyllophoraceae* II. *Meded. Nederl. Mycol. Ver.* 22: 278 pp.

Donk, M.A. (1961). Four new families of hymenomycetes. *Persoonia* 1: 405–407.

Donk, M.A. (1962). Notes on resupinate hymenomycetes VI. *Persoonia* 2: 217–238.

Donk, M.A. (1969). Notes on *Cantharellus* sect. *Leptocantharellus*. *Persoonia* 5: 265–284.
Fiasson, J.-L. (1976). Les caroténoides de *Cantharellus ianthinoxanthus* (R. Maire) Kühner et sa position taxonomique. *C. R. Acad. Sci., Paris* 276, sér. D: 3219–3220.
Ginns, J. (1985). *Hericium* in North America: cultural characteristics and mating behavior. *Can. J. Bot.* 63: 1551–1563.
Ginns, J. & Lefebvre, M.N. (1993) Lignicolous corticioid fungi of North America. *Mycologia Memoir* 19: 247 pp.
Harrison, K.A. (1961). The stipitate hydnums of Nova Scotia. *Canad. Dept. Agric., Ottawa, Res. Branch Publ.* 1099: 60 pp.
Harrison, K.A. (1984). *Creolophus* in North America. *Mycologia* 76: 1121–1123.
Ing, B. (1992). A provisional Red Data List of British fungi. *The Mycologist* 6: 124–128.
Jansen, E. & van Dobben (1987). Is decline of *Cantharellus cibarius* in the Netherlands due to air pollution? *Ambio* 16: 211–213.
Jülich, W. (1981). Higher taxa of basidiomycetes. *Bibl. Mycol.* 85: 485 pp.
Jülich, W. (1984). *Die Nichtblätterpilze, Gallertpilze und Bauchpilze.* Kleine Kryptogamenflora IIb/1. Stuttgart: Gustav Fischer Verlag.
Kuyper, T.W. (1990). Nomenclatural notes on two species of *Cantharellus. Riv. Micol.* 33: 245–249.
Lizon, P. (1995). Macrofungi reported as extinct/missing or threatened with extinction in European Red Data Lists. *Fungi and Conservation Newsletter* 3: 3–4.
Maas Geesteranus, R.A. (1959). The stipitate hydnums of The Netherlands - IV. *Auriscalpium* S. F. Gray, *Hericium* Pers. ex S. F. Gray, *Hydnum* L. ex Fr., and *Sistotrema* Fr. em. Donk. *Persoonia* 1: 115–147.
Maas Geesteranus, R.A. (1962). Hyphal structures in hydnums. *Persoonia* 2: 377–405.
Maas Geesteranus, R.A. (1963). Hyphal structures in hydnums II. *Proc. Koninkl. Nederl. Akad. Wetenschappen* C. 66: 426–457.
Maas Geesteranus, R.A. (1969). Notes on hydnums. VIII. *Proc. Koninkl. Nederl. Akad. Wetenschappen* C. 72: 213–221.
Maas Geesteranus, R.A. (1975). *Die terrestrischen Stachelpilze Europas*. Pp. 1–127, pl. 1–40, figs. 1–587. Amsterdam: North Holland Publishing Co.
Maas Geesteranus, R.A. (1977). Problems around a new *Mucronella*. *Persoonia* 9: 271–274.
Maas Geesteranus, R.A. (1978). Notes on hydnums - XI. *Persoonia* 9: 491–500.
Maas Geesteranus, R.A. & Nannfeldt, J.A. (1969). The genus *Sarcodon* in Sweden in the light of recent investigations. *Svensk Bot. Tidskr.* 63: 401–440.
Marchand, A. (1971). *Champignons du Nord et du Midi* 1: 273 pp, Pl. 1–100. Perpignan: Société Mycologique des Pyrénées Méditerranéennes.
Marchand, A. (1973). *Champignons du Nord et du Midi* 2: 273 pp, Pl. 101–200. Perpignan: Société Mycologique des Pyrénées Méditerranéennes.
Marchand, A. (1976). *Champignons du Nord et du Midi* 4: 261 pp, Pl. 301–400. Perpignan: Société Mycologique des Pyrénées Méditerranéennes.
Monthoux, O. & Röllin, O. (1978). *Cantharellus ianthinoxanthus* et *melanoxeros*, deux espèces distinctes. *Schweiz Zeitschr. Pilzk.* 56: 145–149, pl. 11–6.
Moser, M. & Jülich, W. (1985). *Farbatlas der Basidiomyceten.* (in loose-leaf fascicles, continuing). Stuttgart: Gustav Fischer Verlag
Muskett, A.E. & Malone, J.P. (1980). Catalogue of Irish fungi II: hymenomycetes. *Proc. Royal Irish Acad.* 80(B): 197–276
Nikolaeva, T.A. (1977). Hydnaceae fungi. *Cryptogamous Plants of the USSR* VI, *Fungi* (2).
Nilsson, S. & Persson, O. (1978). *Fungi of Northern Europe* 1. Larger fungi (excluding gill-fungi). Pp. 128. Harmondsworth: Penguin Books Ltd.

Orton, P.D. (1964). Notes on British agarics II. *Notes Roy. Bot. Gard. Edinb.* 26: 43–65, figs. 1–29.

Peck, C.H. (1887). New York species of *Cantharellus. Bull. New York St. Mus.* 1 (2): 1–66.

Persson, O. & Mossberg, B. (1994). *Kantareller.* Pp. 117. Värnamo: Tryckt av Fälths Tryckeri.

Phillips. R. (1981). *Mushrooms and other fungi of Great Britain & Europe.* Pp. 288. London: Pan Books Ltd.

Pyysalo, H. (1979). On the formation of the aroma of some northern mushrooms. *Mushroom Science* 10 (2): 669–675.

Quélet, L. (1872). *Les champignons du Jura et des Vosges* 1: 320 pp. Paris: J. B. Baillière et fils.

Rea, C. (1910). New and rare British fungi. *Trans. Brit. Mycol. Soc.* 3: 220–230, pl. 11–12.

Rea, C. (1922). *British Basidiomycetae, a handbook to the larger British fungi.* Pp. 799. Cambridge: University Press.

Ricek, E.W. (1971). *Cantharellus cibarius* var. *amethysteus* Quél. *Schweiz. Zeitschr. Pilzk.* 49 (6): 71–72, pl.

Ryman, S. & Holmåsen, I. (1984). *Svampar, en fälthandbok.* Pp. 1–718. Stockholm: Interpublishing.

Sowerby, J. (1796–1797). *Coloured Figures of English Fungi or Mushrooms* 1: 120 pl. London.

Stalpers, J.A. (1993). The aphyllophoraceous fungi I: keys to the species of the *Thelephorales. Studies in Mycology* 35: 168 pp.

Stamets, P. (1993). *Growing gourmet & medicinal mushrooms.* Pp. 552. Olympia, Washington: Ten Speed Press.

Vesterholt, J. & Knudsen, H. (1990). *Truede storsvampe i Danmark - en rødliste.*

Wakefield, E.M. & Dennis. R.W.G. (1950). *Common British fungi.* Pp. 1–290, pl. I–CXI. London: P.R. Gawthorn Ltd.

Ying, J., Mao, X., Ma, Q., Zong, Y., & Wen, H. (1987). *Icones of medicinal fungi from China.* Pp. 575. Beijing: Science Press.

INDEX TO TAXA